THE ROLE OF PUBLIC AGENCIES IN FOSTERING NEW TECHNOLOGY AND INNOVATION IN BUILDING

COMMITTEE ON NEW TECHNOLOGY AND
INNOVATION IN BUILDING
Building Research Board
Commission on Engineering and Technical Systems
National Research Council

David R. Dibner
Andrew C. Lemer
Editors

NATIONAL ACADEMY PRESS
WASHINGTON, D.C. 1992

NOTICE: The project that is the subject of this report was approved by the Governing Board of the National Research Council, whose members are drawn from the councils of the National Academy of Sciences, the National Academy of Engineering, and the Institute of Medicine. The members of the committee responsible for the report were chosen for their special competences and with regard for appropriate balance.

This report has been reviewed by a group other than the authors according to procedures approved by a Report Review Committee consisting of members of the National Academy of Sciences, the National Academy of Engineering, and the Institute of Medicine.

The National Academy of Sciences is a private, nonprofit, self-perpetuating society of distinguished scholars engaged in scientific and engineering research, dedicated to the furtherance of science and technology and to their use for the general welfare. Upon the authority of the charter granted to it by the Congress in 1863, the Academy has a mandate that requires it to advise the federal government on scientific and technical matters. Dr. Frank Press is president of the National Academy of Sciences.

The National Academy of Engineering was established in 1964, under the charter of the National Academy of Sciences, as a parallel organization of outstanding engineers. It is autonomous in its administration and in the selection of its members, sharing with the National Academy of Sciences the responsibility for advising the federal government. The National Academy of Engineering also sponsors engineering programs aimed at meeting national needs, encourages education and research, and recognizes the superior achievements of engineers. Dr. Robert M. White is president of the National Academy of Engineering.

The Institute of Medicine was established in 1970 by the National Academy of Sciences to secure the services of eminent members of appropriate professions in the examination of policy matters pertaining to the health of the public. The Institute acts under the responsibility given to the National Academy of Sciences by its congressional charter to be an adviser to the federal government, and upon its own initiative, to identify issues of medical care, research, and education. Dr. Kenneth I. Shine is president of the Institute of Medicine.

The National Research Council was established by the National Academy of Sciences in 1916 to associate the broad community of science and technology with the Academy's purposes of furthering knowledge and of advising the federal government. Functioning in accordance with general policies determined by the Academy, the Council has become the principal operating agency of both the National Academy of Sciences and the National Academy of Engineering in providing services to the government, the public, and the scientific and engineering communities. The Council is administered jointly by both Academies and the Institute of Medicine. Dr. Frank Press and Dr. Robert M. White are chairman and vice chairman, respectively, of the National Research Council.

Funding for the project, initially sponsored by the Department of the Navy, the General Services Administration, and the Department of State, and completed under the Federal Construction Council technical program, was provided through the following agreements between the indicated federal agency and the National Academy of Sciences: Department of the Navy Agreement No. N00025-89-M-0018; Department of State Contracts No. 1030-621218 and 1030-270106; National Science Foundation Grants No. MSS-8902669, under master agreement 8618641; and U.S. Postal Service grant, unnumbered.

Library of Congress Catalog Card Number 92-62883
International Standard Book Number 0-309-04783-8

Additional copies of this report are available from:

National Academy Press
2101 Constitution Avenue, NW
Washington, DC 20418

B 008

Printed in the United States of America

COMMITTEE ON NEW TECHNOLOGY AND INNOVATION IN BUILDING

Chairman

DAVID R. DIBNER, (Retired) Vice President and Principal Architect, Sverdrup Corporation, Arlington, Virginia

Members

ROBERT J. BOEREMA, Construction, Planning and Design Manager, Florida Department of General Services, St. Augustine
LLOYD A. DUSCHA, Consulting Engineer, Reston, Virginia
MARTHA W. GILLILAND, Vice Dean and Assistant Vice President, University of Arizona, Tucson
MICHAEL GREEN, Professor, College of Law, University of Iowa, Iowa City
DONALD E. KASH, Hazel Chair in Public Policy, George Mason University, Fairfax, Virginia
STEPHEN R. LEE, AIA, Instructor of Architecture/Research Associate, Carnegie Mellon University, Pittsburgh, Pennsylvania
ALVIN P. LEHNERD, Vice President Product Development, Steelcase Inc., Grand Rapids, Michigan
PHILIP B. LOVELL, Turner Construction Company, San Jose, California
FRED MOAVENZADEH, George Macomber Professor of Construction Management, and Director of the Center for Construction Research and Education, Massachusetts Institute of Technology, Cambridge
TOM F. PETERS, Director, Institute for the Study of High-Rise Habitat, Lehigh University, Bethlehem, Pennsylvania
ROBERT RATAY, Industry Professor of Civil Engineering, Polytechnic University—Brooklyn and Farmingdale, New York
JOHN W. THOMPSON, Assistant Vice President, Facilities Planning and Development, CIGNA Corporation, Hartford, Connecticut
LOUIS G. TORNATZKY, Scientific Fellow, Office of the President, Industrial Technology Institute, Ann Arbor, Michigan

Federal Construction Council Liaison Representatives

DENNIS BAYON, National Aeronautics and Space Administration, Washington, D.C.
JAMES BINKLEY, U.S. Postal Service, Washington, D. C.
ANDREW CONSTANTARAS, U.S. Army Corps of Engineers, Washington, D.C.
MILON ESSOGLOU, Naval Facilities Engineering Command, Department of the Navy, Alexandria, Virginia
MARVIN GORELICK, U.S. Department of Energy, Washington, D.C.
DAVID M. JONCICH, USA Army Construction Engineering Research Laboratory, Champaign, Illinois
DEBBIE J. LAWRENCE, USA Construction Engineering Research Laboratory, Champaign, Illinois
WILLIAM G. MINER, U.S. Department of State, Washington, D.C.
BILL O'CONNOR, Naval Facilities Engineering Command, Alexandria, Virginia
NOEL RAUFASTE, National Institute of Standards and Technology, Gaithersburg, Maryland
DWAINE WARNE, Public Buildings Service, General Services Administration, Washington, D.C.

Public Facilities Council Liaison Representatives

DALE STANTON HOYLE, Montgomery County Government, Rockville, Maryland
HENRY G. SHIRLEY, Bureau of Capital Outlay Management, Commonwealth of Virginia
DALE STRAIT, Department of General Services, State of Maryland

Project Staff

ANDREW C. LEMER, Director
PATRICIA M. WHOLEY, Staff Associate
SUZETTE CODY, Project Assistant
MARY T. McCORMACK, Project Assistant

BUILDING RESEARCH BOARD
(1991-1992)

Chairman

HAROLD J. PARMELEE, President, Turner Construction Company, New York, New York

Members

RICHARD T. BAUM, (Retired) Partner, Jaros, Baum and Bolles, New York, New York
LYNN S. BEEDLE, University Distinguished Professor of Civil Engineering and Director, Institute for the Study of High-Rise Habitat, Lehigh University, Bethlehem, Pennsylvania
L. GERALD CARLISLE, Secretary-Treasurer, International Union of Bricklayers and Allied Craftsmen, Washington, D.C.
NANCY RUTLEDGE CONNERY, Consultant, Woolwich, Maine
C. CHRISTOPHER DEGENHARDT, Chairman, EDAW, Inc., San Francisco, California
ELISHA C. FREEDMAN, Regional Manager, The Par-Group—Paul A. Reaume, Ltd., Hartford, Connecticut
DONALD G. ISELIN, USN, Retired, Consultant, Santa Barbara, California
FREDERICK KRIMGOLD, Associate Dean for Research and Extension, Virginia Polytechnic Institute and State University, Alexandria
GARY T. MOORE, Professor of Architecture and Director, Wisconsin Space Grant Consortium, University of Wisconsin, Milwaukee
WALTER P. MOORE, President and Chairman of the Board, Walter P. Moore and Associates, Inc., Houston, Texas
J.W. MORRIS, (Retired), President, J. W. Morris Ltd., Arlington, Virginia
BRIAN P. MURPHY, Senior Vice President, Prudential Property Company, Prudential Plaza, Newark, New Jersey
LESLIE E. ROBERTSON, Director, Design and Construction, Leslie E. Robertson Associates, New York, New York
JEROME J. SINCOFF, AIA, President, Hellmuth, Obata & Kassabaum, Inc., St. Louis, Missouri

JAMES E. WOODS, William E. Jamerson Professor of Building Construction, College of Architecture and Urban Studies, Virginia Polytechnic Institute and State University, Blacksburg
APRIL L. YOUNG, CRA Coordinator, First American Metro Corporation, McLean, Virginia

Staff

ANDREW C. LEMER, Director
HENRY A. BORGER, Executive Secretary, Federal Construction Council
PATRICIA M. WHOLEY, Staff Associate
SUZETTE CODY, Project Assistant
LENA B. GRAYSON, Program Assistant
MARY T. McCORMACK, Project Assistant

PREFACE

Charles H. Duell, director of the U.S. Patent Office, urged President McKinley in 1899 to close the agency because "everything that can be invented has been invented." Yet today, the products of technological innovation in many fields are all around us: organ transplants have become almost commonplace in hospitals around the country. A daunting array of new food products with long shelf-lives entices us at supermarkets. Computers that now fit comfortably on one's lap pack more power than the room-sized machines commonly available just two decades ago.

In construction and related industries, change has been less apparent in recent decades. Some observers term the building-related industries "antiquated" and credit them with little real potential for innovation. We and the committee whose work is reported here feel that these critics may be, like Mr. Duell, short-sighted. Nevertheless, evidence suggests that new technology development and innovation in U.S. construction are lagging. At the same time the growing bounty of new products and procedures in other fields suggests that substantial opportunities for building innovation may be emerging.

The agencies of the Federal Construction Council, in asking the Building Research Board to consider what the federal government's role should be in fostering new building technology, have thus raised an issue of much broader consequence than the immediate benefits to these agencies' programs. Innovation leads to improved productivity, better quality, and higher quality of life for individuals and organizations in both private and public sectors.

Pursuit of these benefits is a worthy enterprise. We believe that the federal government indeed does have a crucial role to play, but many other groups in the private and public sectors have important roles as well. We hope through this work to help strengthen the partnership of the many interests needed if our building-related industries are to realize the promise of new technology.

David R. Dibner, *Chairman*
Committee on New Technology and Innovation in Building
Andrew C. Lemer, *Director*
Building Research Board

This study was supported as part of the technical program of the Federal Construction Council (FCC). The FCC is a continuing activity of the Building Research Board, which is a unit of the Commission on Engineering and Technical Systems of the National Research Council (NRC). The purpose of the FCC is to promote cooperation among federal construction agencies and between such agencies and other elements of the building community in addressing technical issues of mutual concern. The FCC program is supported by 18 federal agencies: the Department of the Air Force, the Department of the Army (two agencies), the Department of Commerce, the Department of Energy, the Department of the Interior, the Department of the Navy, the Department of State, the General Services Administration, the National Aeronautics and Space Administration, the National Endowment for the Arts, the National Science Foundation, the U.S. Postal Service, the U.S. Public Health Service, the Smithsonian Institution, and the Department of Veterans Affairs.

The Public Facilities Council (PFC) was formed in 1983 to make available to state and local governments, quasi-governmental authorities, and others, the forum and services of the BRB and NRC to identify technical problems and research needs facing construction administrators and facilities managers. Sponsors of the PFC currently include a score of state and local governments or interstate entities. Funding and participation are typically drawn from the executive office of the jurisdiction responsible for facilities development and management.

Reports resulting from Building Research Board programs are provided free of charge to sponsoring entities.

For information contact:

Director
Building Research Board
National Research Council
2101 Constitution Avenue, N.W.
Washington, D.C. 20418

CONTENTS

EXECUTIVE SUMMARY 1

1. INTRODUCTION 9
The Government Interest, 9
Benefits and Risks of New Technology, 11
Source and Scope of the Study, 12
Structure of the Report, 14

2. NEW BUILDING TECHNOLOGY, INNOVATION, AND GOVERNMENT INTERESTS 17
Public Benefits of Innovation, 18
Government as Purchaser and Funder of Research, 20
Technological Innovation and Public Policy, 22
Impact of the Building Regulatory Process, 25
Government's Current Role in Fostering New Technology, 27

3. NEW TECHNOLOGY AND INNOVATION IN THE U.S. BUILDING-RELATED INDUSTRIES 31
Background, 31
Lack of Data, 34
Attitudes Toward New Building Technologies, 34
Entry Points for Building Innovation, 36
Status of Building Research, 37

Deterrents to Transfer of New Technology, 39
Limited Opportunities for Innovation, 41
Place of Universities and Other Educational Institutions, 44

4. WHAT SHOULD THE ROLE OF THE FEDERAL AGENCIES BE IN FOSTERING NEW BUILDING TECHNOLOGY? 47
Institutional Perspective, 48
Alternative Roles for Government, 50
Evaluation of Strategy Options, 52

5. IMPLEMENTING AN EFFECTIVE ROLE 57
Responsibilities for Taking Action, 57
Institutional Focus Needed, 58
Actions by Facilities Agencies, 60
Enabling Environment, 61
Technology Transfer, 62
Innovation and the Future, 63

APPENDIXES

A. Biographical Sketches of Committee Members and Staff, 65
B. Processes of Technological Innovation, 69
C. Review of Specific Agencies' Stance Toward Building Innovation, 81
D. Federal Laws and Regulations Related to Technological Innovation in Building, 87
E. New Building Technology and Innovation: A Selective Review, 91
F. Tort Law, Deterrence and Innovation: Too Much or Too Little?, 101
G. Points of Entry for Building Innovation, 125

EXECUTIVE SUMMARY

Sixteen agencies comprising the Federal Construction Council (FCC) asked the Building Research Board (BRB) to advise them on factors to be considered in making decisions about the use of new building technology. Some agency officials actively seek to foster development and adoption of new building technology to meet their own needs or to enhance the productivity of U.S. industry. Others suggest that the inherent risks of new technology require that responsible managers be more conservative in the application of new technology in government facilities.

With an aggregate annual construction budget of some $15 billion, should not U.S. government agencies act as a significant market force for beneficial change? On the other hand, as stewards of public assets, can government officials responsible for facilities planning, construction, and management accept the inevitable risks of innovation?

The issues raised by these questions have broad implications. Building-related industries account for a major fraction of the U.S. economy, and their products have pervasive influence on individuals and business. Many observers have expressed concern in recent years that the rate of innovation in these industries has lagged behind other U.S. industries and the building-related industries in other countries. The BRB Committee on New Technology and Innovation in Building, appointed to respond to the FCC request, found itself involved in a more basic discussion of whether there is an appropriate role for the government in fostering the development of new building technology. That is the subject of this report.

The U.S. building-related industries (i.e., design services, construction, building materials and products manufacture, and facilities operations and maintenance) are highly fragmented. There are many distinct segments within these industries, and in many segments there are many small firms with limited geographic scopes of operations. Government agencies' policies on matters of new technology are similarly fragmented. This feature has decisive influence on technological innovation in the building-related industries.

Much of the concern for innovation in these industries has focused on process-related innovation in design and construction. An emphasis on individual construction projects as the basis for analysis has supported increasingly misleading views of the rate of innovation and of the way much new technology is introduced in the industries. Building products and materials manufacturers are an important and neglected source of new technology, a source that is often ignored by studies of innovation in the building-related industries.

Even so, the rates of development of new technology and innovation in the building-related industries have been, for at least the past two decades, lower—and the role of government has been less positive—than they could be, relative to both the industry's potential capabilities and the nation's welfare. There are three primary reasons why the federal government, and government at other levels as well, should play a greater role in fostering new building technology:

1. to achieve better cost (initially and over the course of a building's lifecycle), quality, and performance in government facilities;
2. to enhance the quality of life in the United States generally (and worldwide), through encouraging better cost—initial or lifecycle—quality, and performance in private sector building; and
3. to enhance U.S. industrial competitiveness in international markets.

Government agencies generally encourage new technology through direct purchase for mission-oriented applications or through promulgation of policies aimed at broader social goals (e.g., environmental protection). Both roles are appropriate and necessary but have had only limited impact with respect to building technology. In contrast to many other countries, there is no U.S. government agency with explicit responsibility for representing or encouraging enhancement of the nation's construction industry as a whole. Programs and policies are needed to overcome the industries' and government's fragmentation. Such programs as the percentage-of-construction-cost set-aside for art and the Strategic Highway Research Program are models that have demonstrated success in dealing in the short term with a fragmented environment while building in the long term a coalition of interest.

The committee considered the range of strategic roles that government might play in fostering new building technology, and characterized that range in terms of three broad options representing generally increasing levels of proactive government involvement in the U.S. building-related industries' innovation processes:

1. **Business as usual**—a continuation of relatively conservative current policies and support for new technology—is the baseline against which any recommendations for change must be compared. Actual funding levels may decrease as federal agency construction budgets shift away from military construction toward civilian facilities, and may decline in absolute terms for at least the next decade. State and local governments could, in aggregate, experience substantial growth in construction for the renewal of physical infrastructure, the renovation and expansion of educational facilities, and the accommodation of increased work loads associated with programs formerly under federal administration. Agency research and development activities would be expected to continue at least at current funding levels.

2. **Active mission-oriented technology pursuit** would involve definite action to encourage development of new building technology that can be applied in agencies' own facilities projects. General policies such as required "new technology set-asides" in procurement to fund testing and demonstration, required annual progress reports by each agency to the Office of Science and Technology Policy, or the establishment of presidential awards for new building technology demonstrations could motivate each individual agency's efforts. These efforts might include establishing a centralized "building innovation office" or an interagency coordinating body, or contracting with one of the national laboratories for assistance in identifying promising projects and technologies. A new organization would have certain advantages, but existing agency research and development programs and centers can be expanded and used effectively to identify needs, mobilize the search for solutions, and aid the transfer of emerging technology into practice.

3. **Increased federal research, development, and demonstration support** would be targeted to benefit state and local governments and the private sector, as well as federal agencies themselves. The National Science Foundation, the Advanced Technology Program of the National Institute of Standards and Technology, and the Strategic Highway Research Program, all congressionally funded, and the U.S. Army's CPAR[1] program are potential models for how this support might be provided. Tax disincentives that discourage investment could be reduced and actions taken to limit liability

[1]Construction Productivity Advancement Research.

exposure and awards, limit the diversity and restriction of building regulations, and generally reduce regulatory and administrative impediments to the application and commercialization of new technology, consistent with the protection of life safety, health, and environmental integrity.

The committee evaluated alternative strategies in terms of their potential consequence for cost and performance of government facilities, U.S. quality of life overall, and U.S. international competitiveness. The committee concluded that **the federal government's position as a major provider and user of facilities effectively imposes on individual agencies a responsibility to pursue innovation and foster new technology.** A government-wide "business as usual" strategy is therefore not appropriate. Because agencies differ in their missions, resources, and needs, no one strategy is appropriate throughout government. Rather, strategy for fostering new building technology should be tailored to each agency, generally incorporating elements at several levels of proactive involvement: (1) federal agencies' encouraging applications of new technology for their own projects, (2) multiagency activities to enhance the pursuit and effective transfer of new technology to the U.S. private sector, and (3) generally increased support for targeted efforts to develop new technologies in specific areas, perhaps through existing and new university- and industry-based "centers of excellence." All agencies with interest in facilities and their construction would have some part to play in implementing this composite strategy.

An institutional mechanism to focus attention on technology in the building-related industries is needed. There is no single agency or program in the federal government with comprehensive responsibility for dealing with issues of construction and facilities. At the same time, many agencies that build facilities can benefit directly from innovation in these facilities. These agencies have separate missions, operate independently, and are in some senses competitive in their traditions and administrative procedures. For these reasons, the committee concluded that the institutional focus needed is outside any single construction agency's existing facility program. Responsibility and resources for coordinating government building-related innovation strategy, for taking positive leadership in implementing this strategy, for fostering action by construction agencies, and for evaluating progress, should be assigned to the agency or office in which this focus is established.

Existing agencies might assume these functions (e.g., within the Department of Commerce or Department of Housing and Urban Development), but there are models for creating new organizations (e.g., the Council on Environmental Quality). In either case, a broad scope for this organization would include support of technology development and demonstrations in government facilities, funded by a percentage set-aside or "tax" on all agencies' construction appropriations. Agencies would then compete for support of their specific projects. Such a commission or other focus should be established with a limited life, perhaps five years, but might be renewed if progress is demonstrated clearly. The evaluation and information dissemination functions, however, should continue.

Mission-oriented agencies responsible for design and procurement of facilities should adopt measures to encourage designers and constructors to propose and perfect cost-effective new technologies, such as

- using new technology as a selection factor, by giving particular credit in design and construction procurement to offerers who propose to apply potentially cost-effective new technology that has been developed at least to the stage of prototype application;
- using performance specifications in construction procurements to permit offerers to propose new technologies that may not meet more traditional (i.e., prescriptive) standard specifications; and
- using such procurement mechanisms as design-build or build-operate-transfer to promote increased integration in the delivery process, both to permit innovators to apply new technology at any stage of the facility life cycle and to assist the innovator to recapture benefits of innovation.

Budgets must be made available to pay for added costs of planning and design analyses that may be required and for evaluation of completed installations. Responsible officials must have adequate time to oversee projects on which new technology is being applied and to assess and document results. In addition, **senior agency and congressional officials must accept that some new technologies may not perform as expected**. Systems are needed to provide careful review and to offer insurance and indemnification for both providers and users of new technology. More importantly, support must be reliably sustained long enough to permit investments in research, development, and technology transfer to yield results.

To enhance technology transfer, agencies should reward efforts to innovate, by establishing programs to promote projects showcasing new technology or design competitions based on applications of new technology. Award programs (similar perhaps to the Presidential Design Awards or the Malcolm Baldridge National Quality Award) for building innovation would highlight the contribution of these efforts to agency effectiveness. Further, industry should be intimately involved in the direction of government spending to encourage high-priority research and development activities.

Although major growth in the estimated $200 million to $230 million spent annually, for federally supported building research, development, and demonstration is unlikely in today's fiscal climate, the scale of the market suggests that relatively modest increases of 10 to 15 percent could produce significant results, particularly if the spending is matched by private sector contributions and concentrated in a relatively few institutions. **The building-related industries should petition Congress to establish a program of integrated research, development, and incubation of new building technology, to support establishment or continuation of several "building technology centers of excellence."**

Changing international relationships may reduce the scale of military expenditures, making resources available, in the United States and elsewhere, to seek improvements in our quality of life. New technologies are emerging that offer greater opportunities for enhanced service, greater efficiencies, and protection of natural environmental resources. The growth of global markets can enhance productivity and the dissemination of building technology improvements. However, over the shorter term, the risks and inevitable discomforts of change must be managed rather than permitted to block progress. The U.S. building-related industries are being called on to evolve under conditions of uncertainty. Government agencies, as both users of the industries' products and instruments of national policy, have an important role to play in fostering new technology and the innovation on which the future of these industries depend.

THE ROLE OF PUBLIC AGENCIES IN FOSTERING NEW TECHNOLOGY AND INNOVATION IN BUILDING

1

INTRODUCTION

The U.S. building-related industries—planning, design, construction, building materials and products, equipment manufacturing, facilities operations and maintenance—are an important part of the nation's economy, whose products touch the lives of everyone. Construction alone typically employs some 5 million to 6 million people and, with an annual output of some $440 billion (in 1990), accounts for approximately 7 to 9 percent of gross domestic product. New building technology and subsequent innovation can, in these construction-related industries, have a major impact on productivity, not only within the industries themselves but also for the myriad social and economic activities that constructed facilities shelter and serve.

THE GOVERNMENT INTEREST

The federal government spends some $15 billion annually for the design and construction of buildings and other facilities. State and local governments, in aggregate, account for perhaps an additional $50 billion each year. Billions of dollars more are spent to operate (e.g., heat, cool, and illuminate) and maintain government facilities.

Government agencies thus have a substantial stake in new technology and the rate of innovation in building-related industries: design services, construction, building materials and products manufacturing, and facilities operation and maintenance. These agencies—and the taxpaying public—stand to gain directly from the development and use of improved building products and processes.

Some agency officials would seek to foster the development and adoption of new building technology to meet their own needs or to enhance the productivity of U.S. industry, but feel that slow domestic innovation and restrictions on the purchase of foreign goods and services (i.e., "Buy American" and other programs) combine to place government programs at a particular disadvantage in using new technology when renovating aging facilities or developing new ones. Other officials suggest that although new technology may present opportunities for improved performance and productivity, the risks of applying technology that is not yet widely employed require that responsible government managers use proven methods and materials. In addition, agency officials must answer to the public and its elected representatives, and may hesitate to accept risks that cannot be justified by clear opportunities for financial returns.

As a result, agencies contracting for the design and construction of public buildings and other facilities have often adopted conservative policies toward the application of new building technology. However, many government officials claim both the willingness and, if government policy is supportive, the ability to adopt new technology that promises solid payoffs in enhanced economy or performance of public facilities.

There are also those who assert that government agencies, and federal agencies especially, should do even more. These proponents of a more aggressive policy refer to seemingly low rates of innovation in the building-related industries—particularly in comparison to rapid advances in electronics, biotechnology, communications, and other fields—and suggest that government should actively foster development of new building technology. They argue that government agencies have unique ability to share the financial risks associated with trying new building materials, equipment, designs, or techniques, and in doing so would provide leadership and enhance innovation rates. They argue that conservative policies toward new technology hinder the government's and the private sector's ability to gain early benefit from technological innovation and, consequently, that the nation misses opportunities for improved productivity and quality of life.

As the latter point suggests, government's concern for innovation extends beyond that of the government as purchaser. The president's 1990 statement of U.S. technology policy (Office of Science and Technology Policy, 1990) asserted the need for our society to focus on ensuring the translation of technology into "timely, cost effective, high quality manufactured products" and "a legal and regulatory environment that provides stability for innovation and does not contain unnecessary barriers to private investments in R&D and domestic production." **Government plays a critical role in establishing a favorable environment for innovation through private industry's research and development activities and the aggressive pursuit of commercial applications of technologies resulting from these activities.** In the case of some technologies such as solid-state electronics and new materials, the government has played a very active role in funding both basic research and product development, primarily for defense applications. Proponents argue that this role should be pursued in the case of building technology.

BENEFITS AND RISKS OF NEW TECHNOLOGY

Controversy over government policies toward new building technology reflects the general conflict between the rewards of enhanced performance and reduced cost that new technology can offer, and the risk that this new technology may not work. With building technology, this risk can involve not only costs to repair, replace, or make do with the facility or component that has failed to perform as anticipated, but also the safety of construction workers and facility users. Because of the long service lives expected of constructed facilities, the benefits and risks of new technology can be long lasting as well.

In one example of the controversy, the inventor of a new base isolation system[2] proposed to at least two federal agencies that they use his devices in the seismic upgrading of their structures. Although similar systems have been used, experience with their performance in actual earthquakes is very limited, especially in the United States. The particular device in question was completely untried. Agency officials declined to adopt the device and subsequently had to explain their decision to congressional inquirers.

In another example, an agency permitted the use of an experimental additive to concrete to enhance that material's resistance to the chemical and physical attack of seawater, hoping to increase the service lifetime of the structure and reduce the costs of maintenance. The experiment failed, and the agency was faced with costly repairs.

Many times, the adoption of new technology is successful. For example, single-ply roofing[3] was developed initially in Europe and advanced by U.S. government and private sector research. With new high-performance membranes and effective seam-sealing techniques, single-ply systems have in slightly more than one decade captured approximately 35 percent of the U.S. nonresidential roofing market. Recycling of highway pavements, roller-

---

[2]Base isolation devices are designed to isolate a building's superstructure from the underlying soil and rock through a foundation system that attenuates seismic force transmission, thereby potentially reducing earthquake-caused damage to the structure, its contents, and its occupants.

[3]This is in contrast to the multiple layers of material built up to shape and reseal older conventional roofs on large buildings.

compacted concrete, and low-energy-loss windows are other new technologies developed in the past several decades that have become common because they offer cost-effective and safe improvements in performance.

However, special leadership is sometimes needed to motivate the adoption of new technology. For example, the Urban Mass Transportation Administration sponsored the first U.S. use of precast concrete segmented tunnel liners on one part of the Baltimore subway system—and accepted the associated risk—years after such liners were widely used in Europe.

Such cases illustrate the broad conflict that anyone responsible for selecting and using new building technology must face. New technology can yield real benefits in time and cost savings and improved performance, but in view of limited budgets and possible safety hazards, the decision maker must weigh carefully the possible consequences of adopting any new material or system.

The decision can be especially difficult for government agency officials. These officials must consider not only the cost and safety implications of their decisions, but also the impact of other public policies that may be unrelated to the facilities themselves. Valid differences in priority among conflicting public policies may lead to decisions for or against the adoption of new technology in any particular instance, and those seeking government adoption of their inventions may understandably disagree with agency decisions. However, agency decision makers may be called on to justify their judgments to other executive or legislative officials.

SOURCE AND SCOPE OF THE STUDY

Faced with these concerns, the agencies of the Federal Construction Council (FCC)[4] asked the Building Research Board (BRB) to advise them on factors to be considered in making decisions about the use of new building technology. Recognizing that the users and benefits of new technology are not restricted to federal government, the state and local government agencies of the Public Facilities Council (PFC)[5] also participated in the study.

---

[4]The FCC is a group of 18 federal agencies with interest in facilities and construction. These agencies have combined construction budgets exceeding $7 billion annually.

[5]The PFC is a group of some 20 state and local government agencies responsible for the design, procurement, and management of public facilities in their jurisdictions. These agencies sponsor and participate in selected BRB activities, in a manner similar to the agencies of the FCC.

To respond to these agencies' request, the BRB established the Committee on New Technology and Innovation in Building to study new technology and innovation in the U.S. building industries.[6] Meeting over the course of about 14 months, the committee undertook to identify key issues and develop its recommendations. Liaison representatives from sponsoring government agencies and other invited participants worked with the committee. This report is the product of the committee's work.

The study began with the topic of how government agencies should consider new technology in their building programs, but also looked beyond to the larger concerns of the building-related industries. It is said that we live in a technological age, and we have come to accept and to expect a steady stream of new products and services based on applications of science to practical purposes. In comparison to great strides being made in electronics, medicine, or biotechnology, many observers contend that innovation in buildings and construction—in the United States of the late twentieth century, at least—is lagging.

However, data and careful analyses to support such contentions are sparse. Studies made by the BRB and others generally show that aggregate spending by industry and government on research and development in building-related fields—at about 0.4 percent of annual construction output—is well below levels that the industry's size and importance in the nation's economy warrant, and proportionally less than other industrialized countries (BRB, 1988). However, R&D figures are, at best, a proxy indicator of technology development. Research and development does not lead inevitably to innovation, nor does the absence of explicitly identifiable R&D activity necessarily mean that innovation will not occur. The term R&D is used to refer to a range of activities that researchers and research managers may identify more explicitly and distinctly with such terms as basic research, applied research, technology transfer, testing and evaluation, or others.

Various aspects of the U.S. building-related industries deter innovation: many smaller firms operating in narrow geographic areas, a complex regulatory framework of building codes and liability litigation, and government policies that seem to discourage the longer-term view that business needs to realize the payoffs of developing new technology. Yet none of these characteristics has been shown to be decisive in limiting opportunities for new building technology.

Past innovation in building is poorly documented, and there are no generally accepted bases for measuring this innovation. Although interindustry and international comparisons do suggest that the U.S. industry is lagging (e.g., Civil Engineering Research Foundation, 1991), it is difficult to assert with

---

[6]Brief biographical sketches of committee members and staff are presented in Appendix A.

confidence that the rates of U.S. building innovation are low by any absolute measure.

Nevertheless, regardless of current or historic experience, increased rates of new building technology development and innovation can yield broad benefits to the industry and the nation. Some observers argue that higher rates of innovation are a competitive necessity (e.g., Tatum, 1989), and committee members familiar with Japanese and European construction industries were inclined to agree.

Thus, the committee addressed itself to the broad concern of whether specific government actions are warranted as a cost-effective way of motivating technological innovation in the U.S. building-related industries, innovation that can improve performance and lower costs of buildings and construction for both the private and the public sectors.

The committee's primary goal was to develop practical recommendations regarding government's role—as purchaser of goods and services for its own use and as an instrument of broader public policies—in fostering, developing, applying, and transferring new technology and innovation in the design, construction, maintenance, and rehabilitation of buildings and other facilities. The committee focused its discussion particularly on the following issues:

- characteristics, current status, and rates of occurrence of building innovation, in the United States and other countries, that suggest the need for government policy or action;
- how characteristics of the U.S. market influence building innovation and perhaps thereby limit or direct government policy or action;
- whether current rates of innovation and new technology development in building are affecting the ability of the U.S. building industry to support a high domestic quality of life and compete internationally, to an extent that warrants government attention;
- whether the roles played by private industry and universities, as well as government, in building innovation could be made more effective through changes in government policies or programs;
- actions that could be taken by government, private industry, or others to foster innovation in building; and
- procedures for implementing the committee's recommendations.

STRUCTURE OF THE REPORT

This report documents the committee's deliberations, conclusions, and recommendations. Chapter 2 reviews selected research and analyses on technological innovation that formed a background for much of the committee's discussion and considers in general terms the interests of government at national, state, and local levels in fostering new technology. These interests are reflected in a range of possible strategies that provided a basis for testing the committee's conclusions and developing recommendations.

Supporting Chapter 2 are three appendixes. Appendix B reviews briefly key points from the vast literature on technological innovation. Appendix C reviews the current practices of several government agencies. Appendix D presents a list of principal federal laws and regulations with direct relevance to this study.

A thorough review of innovation in the building-related industries was beyond the scope and resources of the committee's study. Chapter 3 draws largely on the committee's experience to assess the nature and status of U.S. building technology, how innovation in the U.S. building-related industries is influenced by the structure and operating environment of these industries, and the importance of innovation to these industries and their clients and customers.

Chapter 3 is supplemented by three appendixes. Appendix E summarizes background discussions on technological progress and innovation in the building-related industries, which formed the basis for the assessment in Chapter 3. Appendix F is a review of U.S. tort law and its influence on innovation in the construction-related industries. Appendix G discusses in some detail the various ways in which new technology can become innovation.

Chapter 4 presents the committee's assessment of the merits of strategic roles that the U.S. government might play in fostering new building technology and innovation. The result of the committee's deliberations is a proposal for change, at both the policy and the operating levels, involving federal agencies' encouraging applications of new technology for their own projects, government activities to enhance the effective transfer of such technology to the U.S. private sector, and increased support for targeted efforts to develop new technologies in specific areas, perhaps through existing and new university- and industry-based centers for the development and demonstration of new technology.

Such change does not come easily, and will require action by agency officials and encouragement by congressional leaders. Chapter 5 presents the committee's specific recommendations for actions to implement its proposals.

REFERENCES

Building Research Board (BRB). 1988. Building for Tomorrow: Global Enterprise and the U.S. Construction Industry. Washington, D.C.: National Academy Press.

Civil Engineering Research Foundation and Japan International Research Task Force. 1991. Transferring Research into Practice: Lessons from Japan's Construction Industry. Washington, D.C.

Office of Science and Technology Policy, Executive Office of the President. 1990. U.S. Technology Policy. Washington, D.C.

Tatum, C. B. 1989. Organizing to increase innovation in construction firms. Journal of Construction Engineering and Management. 115(4):602-617.

2

NEW BUILDING TECHNOLOGY, INNOVATION, AND GOVERNMENT INTERESTS

There are at least three key reasons why government, in general, and the federal government in particular, might take special interest in new technology and innovation in building:

1. to achieve an appropriate balance of cost (initial or over the course of a building's life cycle), quality, and performance in government facilities;
2. to enhance quality of life in the United States generally (with worldwide benefits as well) by encouraging better cost—initial or life cycle—quality, and performance in private sector building; and
3. to enhance the productivity and commercial success of U.S. construction-related industries in domestic and overseas markets.

Moderating such interests is government's responsibility to protect its citizens and the national interest by avoiding harmful technologies and conserving scarce resources. This responsibility is reflected in a variety of building codes and regulations, as well as broader laws and review procedures.

Government agencies can seek to foster—or at least support—the development and adoption of new technology through a variety of actions.[7] Aggressive adoption of new ideas in the commercial marketplace, direct

[7]Appendix B presents a summary of activities by agencies sponsoring this study.

purchase of new technologies that may not yet meet generally applied standards for acceptance in the private sector, and solicitation of technologies that meet new higher standards set by government are mechanisms typically used to encourage private sector activities leading to innovation. Encouragement of joint public and private efforts to develop new technologies for specific applications, and financial support for research and development in research centers, represent more directed involvement in the new technology development process. Government can also provide legal protection, risk sharing, indemnification, and other indirect methods of increasing private incentives to innovate. Support for education—in universities and for practitioners in all phases of building—enhances both the nation's research capability and the market's propensity to test and adopt potentially valuable new ideas. Various agencies have, from time to time, pursued all of these means, sometimes as a matter of agency policy and sometimes motivated by legislative requirements.[8]

PUBLIC BENEFITS OF INNOVATION

Economists tell us that one of the basic determinants of growth in the economy as a whole is growth in productivity, (i.e., improvements in the efficiency of the economic engine of an area or nation). In turn, technological change and the investment embodying that change, employed by properly trained people, are the keys to productivity growth and rising standards of living (Landau and Hatsopoulos, 1986). One historian, reflecting on this relationship of technological creativity and economic progress, has suggested that the difference between rich nations and poor nations is simply that the former produce more goods and services, because their technology—their ability to control and manipulate nature and people for productive ends—is superior (Mokyr, 1990).

Technological change is any alteration in the production process to increase efficiency; it may result from the application of entirely new information or the diffusion of existing information to new users. Such change springs largely from innovation.

Innovation is an abstract concept, having to do with putting new ideas into practice, but specific products and procedures reflect the result of innovation and are viewed by most people as the embodiment of innovation[9] (see box).

---

[8]Appendix C presents an overview of key federal laws and regulations.

[9]The literature on technological innovation and its history, economics, sociology, and political science aspects is vast. Appendix B presents a brief discussion of key points and definitions from this literature that the committee considered.

THINKING ABOUT NEW TECHNOLOGY AND INNOVATION

Innovation, the introduction of a new idea, entails both *production* and *transfer* of new information to people who can use that information to solve problems, to see the world in a new way, or to enhance their efficiency, effectiveness, or living quality. Technological innovation may be a new product or process of production; a substitution of a cheaper material in an otherwise unaltered product; or the reorganization of production, internal functions, or distribution arrangements, leading to increased efficiency, better support for a given product, or lower costs. Many of the construction industry's technologies involve combinations of hardware and software.

Technological innovation can also be an improvement in the ways of making or doing innovation. Industrial research facilities such as the telephone industry's Bell Labs have been recognized as major contributors to innovation in electronics.

New technology that is not put to productive use is not innovation. At the same time, even technology that is well known and widely used in some industries or places may be new and innovative in a different setting.

Successful new technology and innovation tend to be inspired primarily by practical needs. Technological innovation may also be initiated by scientific invention—new discoveries and developments—but "market pull" is felt widely to be more influential than "technology push" as a force for innovation.

Although innovation improves productivity, innovative individuals or groups are not necessarily more productive. Nevertheless, the absence of innovative adaptation to a rapidly changing environment is a generally reliable indicator of future decline and possible extinction for economic enterprises as well as biological species.

However, the distinction between the invention of a new product or process and subsequent innovation is important, because **putting new ideas (i.e., new products or procedures) into practice determines if innovation has occurred.** In addition, a distinction should be made between adoption of new ideas and achieving improved effectiveness (i.e., greater speed, profitability, competitiveness, quality, safety, or some other measure of success). Innovation is important not for its own sake, but rather for the benefits it can bring to the individuals, organizations, and societies that use it (Tornatzky, et al., 1990).

Innovation can occur in all stages of a facility's life cycle (from programming, planning, design, and construction, through repair and maintenance), and in the software (e.g., design procedures, contracting, administration, management) as well as the hardware of a building. Much of the innovation in construction occurs in response to unique problems encountered on an individual project, but much also is embodied in equipment, materials, and products that emerge from conventional manufacturing operations and are to be used in the design office or on the construction site.

As discussed in Chapter 3, new technology is developed throughout the construction-related industries, and even casual observation reveals that the past several decades have brought many beneficial changes in the processes and products of building. "Beneficial" may be defined in terms of reduced time, cost, or hazard in planning, design, and construction; improved performance of the product (i.e., the constructed facility or its elements); or all of these. Performance encompasses durability, safety, and a range of other factors—both qualitative and quantitative—that have direct or indirect impact on the owners and users of facilities.

GOVERNMENT AS PURCHASER AND FUNDER OF RESEARCH

In general terms, government's influence on building technology has a long history. According to historians, what technological progress there was in the classical world, especially in Roman times, served public rather than private purposes (Mokyr, 1990). Roman leaders gained popularity and power by carrying out successful public works. The Rome of A.D. 100 is said to have had better paved streets, water supply, and fire protection than the capitals of civilized Europe in 1800. Supplying Rome with water was begun by Appius Claudius in 312 B.C. and the system reached unprecedented complexity in the first and second centuries A.D. Sewage and garbage disposal systems were also highly developed. Cement masonry, observed to have occurred earliest in Asia Minor and reported in North Africa by Pliny, is viewed by some as the only great discovery that can be ascribed to the Romans, who vastly improved its use and quality control. Commissioning of the Thames tunnel in nineteenth century England fostered major advances in underground construction,[10] and London's Crystal Palace, a landmark of Victorian England, did likewise for flat-glass manufacturing.

Since World War II, the U.S. government has become an active sponsor of research and development, which has in turn led to much innovation. By one estimate, the federal government provides about 46 percent of all funds for R&D spending in the United States (National Science Foundation, 1990). These funds

---

[10]The Brunels' success with the shield tunneling method was considered a triumph and helped to consolidate Marc Isambard Brunel's reputation.

are expended at government laboratories, universities, quasi-governmental laboratories,[11] and private research organizations.

Much of this spending has been directed toward military ends. Weapons systems have unquestionably become more sophisticated as a result of this concentrated R&D effort.

However, R&D activity does not necessarily lead to innovation. Federal R&D programs in nondefense areas have often been criticized for underestimating the challenge in transferring new information into practice.

This transfer can generally occur in two principal ways. One is through publication of reports, organization of workshops, and other noncommercial mechanisms for information exchange. The other is through commercialization of the new product or process via patenting and licensing, new venture sponsorship, or other means for bringing new technology to the marketplace. Government programs have historically focused primarily on the former approach, although a number of federal programs initiated in the 1980s have been intended to shift the emphasis toward commercialization (see Appendix C).

One of the more aggressively pursued programs has been the U.S. Army Corps of Engineers' still young Construction Productivity Advancement Research (CPAR) program. CPAR seeks to demonstrate how the research payoff can be improved, by allowing the government's private research partners to make profits from the development of ideas produced under the program.

Frequently the problems of commercialization have involved the substantial investment required to develop new technology from research demonstration to marketable product. For example, the Solar Energy Research Institute[12] was unable to interest manufacturers or trade associations in new vacuum glass and electrochromic windows developed during the mid-1980s, because of this investment cost (*Inc. Magazine*, 1987). However, another reviewer of federally supported R&D to improve energy efficiency in the building sector suggested that resulting commercially viable innovations may have been brought into use years sooner than would have been achieved by the private sector alone (Geller et al., 1987).

New technologies developed for application in government facilities often have a better chance for more widespread use. The U.S. Army's Construction Engineering Research Laboratory, for example, is able to show returns on investment (measured by life-cycle cost savings on government facilities) for 21

---

[11]Such institutions as the Argonne National Laboratories and the Mitre Corporation were founded as adjuncts to government agencies and still enjoy a special status in their access to government research and development funding.

[12]This research organization, located in Colorado, was established initially by the U.S. Department of Energy.

research projects, conducted over the course of a decade, ranging from 4:1 to more than 100:1.

As already noted, roughly half of all estimated U.S. Research and Development is federally funded, but only a small portion of this funding is devoted to construction. A 1985 Building Research Board (BRB) study estimated that federal agencies, acting as both sponsors of research and users of research results, spent some $200 million annually on construction research (i.e., about 16 percent of all research in the field). BRB staff estimates suggest that amount may have increased to approximately $230 million (in current dollars, an increase of about 15 percent) in fiscal year 1991. Total direct program spending by federal agencies for new construction, amounting to nearly $13 billion in 1985, grew to $16 billion over the same period (approximately a 23 percent increase) (BRB, 1988; MacAuley, 1990). The committee was unable to determine the impact of this apparently declining federal commitment to building-related research on rates of new technology development and innovation.

Other areas of federal R&D spending might yield new technology that could be transferred to the building-related industries. Such "harvesting" in electronics and materials manufacturing has yielded a variety of new products, from air traffic control devices to children's games. The committee noted that there is no government agency or organization responsible for this task, which requires imagination as well as stamina. The results to be harvested are often difficult to find within a daunting array of government programs and agency operations.

TECHNOLOGICAL INNOVATION AND PUBLIC POLICY

For a government concerned with the well-being of its citizens, technological growth—and, in turn, innovation and new technology—necessarily become matters of public policy. Experience suggests that in certain technical areas, the free market system, operating undirected, is unlikely to produce technological innovation at rates that are achievable and desirable from the point of view of society as a whole (Mokyr, 1990). Several factors account for this less-than-optimum performance:

1. The costs or disbenefits of adopting new technology are both private (the inventor) and public (society as a whole). However, the latter are often poorly recognized in the allocation of costs and rewards for innovation in the free market system. Hence, the public has been asked to bear the burden of risk or adverse impact of past innovation and may resist new technology when it does not understand the benefits.

2. Government and the rule of law are required to protect inventors' rights and allow innovators to reap the benefits that offset the costs and risks of innovation. When patents cannot be granted or enforced, government is sometimes called on to use direct funding of R&D costs or other ways to compensate or encourage inventors.

3. Market size and integration influence both the generation and the spread of new information. Government can facilitate the introduction of new technology in small or poorly integrated markets.

Government involvement with issues of technological innovation is endorsed and mandated by a variety of legislation (refer to Appendix C). Much of this legislation has been associated with defense or industrial technology and applies to building technology only in general terms. Some programs, such as the Small Business Innovation Research program that is implemented through specialized agencies such as the Department of Transportation and the Department of Defense, may include building technology as an area of specific interest, although the overall program does not.

The Army's CPAR program, the Department of Energy's Building Technologies programs, and the activities of the National Science Foundation (NSF) are the primary examples of federal activities intended specifically to advance building technology. These programs act through contracts and grants for specific studies, typically for projects of relatively short duration. Programs in some states, such as Pennsylvania's Franklin Partnership, may also support building technology, although state programs generally are oriented toward activities viewed more as "high tech" and likely to enhance opportunities for employment growth in new industries.

Procurement methods that encourage new technology have sometimes been used by government agencies.[13] The development of performance specifications and the solicitation of design-build proposals are among these methods. Both state (e.g., Florida) and federal (e.g., Corps of Engineers and Naval Facilities Engineering Command) agencies have had some success in bringing designer and constructor together early in the development process, a technique sometimes termed "partnering."

Another proposal that has received some support is the establishment of research centers focused on particular areas of building science and technology. NSF has provided funding for centers devoted to cement technology (at Northwestern University), earthquake engineering (the State University of New York at Buffalo), and studies of large-scale structures (Lehigh University). The

---

[13]Some observers feel that government procurement practices represent a serious deterrence to innovation. However, this observation now is the subject of some dispute and may be the basis of a future BRB study.

Department of Transportation administers a congressionally established set of 10 regional transportation "centers of excellence," and the Army's Construction Engineering Research Center maintains a close working relationship with its neighbor, the University of Illinois.

Besides direct purchase of new technology and funding of R&D, government influences innovation and new technology indirectly through regulatory policy or tax policy. New regulations may create a market for new technology, as is the case with building seismic safety devices, the increased use of residential smoke detectors, and changes in the designs of construction equipment to meet federal safety standards. The potential ban on chlorofluorocarbon refrigerants has spurred research and development to find alternatives.

Tax credits, immediate expensing of R&D expenditures (i.e., rather than having to amortize expenditures over the life of the resulting products), and other tax-based incentives for private research and development have encouraged innovation in a number of fields, including construction-related industries. Sometimes the incentives work indirectly. For example, tax advantages for the renovation of historic structures led to rapid growth in this market in the 1970s and 1980s, supporting in turn the development of many new construction procedures and products suited to the particular problems of preserving and rehabilitating aging materials. Incentives for energy conservation have had a more direct but nevertheless similar impact on insulation technology.

Policymakers in the 1970s argued that the U.S. government, as a whole, was funding a substantial amount of research and development activity that could be used more effectively throughout government and have commercial application. The Federal Laboratory Consortium (FLC) was established in 1974 to provide a basic link between government laboratories and potential users of government-developed technologies. This cooperative network was supplemented when Congress enacted the Stevenson-Wydler Technology Innovation Act of 1980[14] to foster technology transfer from government to the private sector. This act established technology transfer as a mission requirement of each federal laboratory, and required each lab and R&D center to cooperate in programs to advance technology transfer. While it established a technology transfer mission, this legislation's effectiveness was hampered by ambiguities regarding licensing, procedures for undertaking cooperative research with industry, and others.

The Stevenson-Wydler Act, which referred to the role of the FLC, was supplemented by the Federal Technology Transfer Act of 1986,[15] which provided a formal charter and limited funding for the FLC's activities. The

---

[14]PL 96-480, as amended, 15 U.S.C. 3710a et seq.

[15]PL 99-502.

latter act was further strengthened by a 1987 executive order and has undergone other limited, more recent, modifications.

Each federal laboratory and R&D center is generally responsible for developing its own program for technology transfer. However, these research institutions were created to meet the needs of government agencies rather than to assist industry. As a recent study for the U.S. House of Representatives noted, "technology transfer roles and responsibilities are imposed upon a system which is made up of participants not accustomed to working together" (Congressional Research Service, 1991). A study by the General Accounting Office (1991) found that the laboratories differ substantially in their efforts to fulfill this responsibility, and nearly half have no structured program for technology transfer.

On the whole, the committee was unconvinced that substantial progress has been made, in technology transfer from federal laboratory R&D, but it felt that opportunities exist, particularly with regard to such areas as building materials and products. Nevertheless, some committee members were skeptical, questioning whether important new technology has been produced that has yet to be transferred. As is discussed in Chapter 5, this is an area that warrants further investigation.

Finally, government facilitates development and diffusion of new ideas by supporting education and communication in the building professions. Educational and training scholarships and fellowships provided through the National Science Foundation serve such a purpose, although the level of spending has fallen substantially in recent years. State and federal cooperation in the highway program has helped support the nation's transportation research activities, which have produced innovations in several areas of construction.

IMPACT OF THE BUILDING REGULATORY PROCESS

The construction of buildings is regulated by the government for the purpose of protecting and ensuring the health, safety, and welfare of the building's occupants. The authority to administer and enforce building regulations lies with the states, although in actual practice, local municipalities have typically been given this responsibility, without general statewide regulation. In fact, there are more than 44,000 code-enforcing units in the United States, and only 36 states have maintained preemptive control at the state level. This fragmented administration leads to difficulties for all parties in the construction process, and is often cited as a disincentive to intermunicipality and interstate construction activity. Today, in the United States, three primary model codes are available for adoption with or without modification by local code-enforcing units:[16]

1. Basic National Building Codes (NBC or BOCA code) by the Building Officials and Code Administrators International (BOCA),
2. Uniform Building Codes (UBC) by the International Conference of Building Officials (ICBO), and
3. Standard Building Codes (SBC) by the Southern Building Code Congress International (SBCCI).

The NBC code is prevalent in the Northeast, UBC in the West, and SBC in the South. These building codes set forth definitions, standards, and regulations governing occupancy classifications, building types, egress, fire resistance, and structural requirements. The remaining essential building systems are governed by mechanical, plumbing and sprinkler, electrical, and accessibility codes published by a variety of agencies. These agencies include the International Association of Plumbing and Mechanical Officials and the National Fire Protection Association, as well as BOCA, ICBO, and SBCCI.

In many topic areas, the model codes reference or adapt documents provided by standards-writing agencies such as the American National Standards Institute; the American Society of Heating, Refrigerating and Air-Conditioning Engineers; the American Society of Testing and Materials; and the Underwriters Laboratory. Many other groups are active in narrow subareas.

Many efforts to consolidate the plethora of building codes into either statewide or national codes, since the early 1900s, have failed to have major impact. However, with the post-World War II construction boom, interest began anew, and in 1966, under the leadership of the National Bureau of Standards (now the National Institute of Standards and Technology), the National Conference of States on Building Codes and Standards (NCSBCS) was formed to enhance cooperation between states and to assist states with the development of statewide code control to improve the regulatory process. NCSBCS continues to be active today and works closely with the systems-built housing industry to assist its interstate regulatory procedures. Nevertheless, the government regulatory process in the private sector is still generally complex and a real barrier to innovation.

In addition to these regulatory controls, the courts play an influential role. Issues of liability for loss and damage associated with new technologies that fail to perform as expected or that have unanticipated effects have been a central

[16]This is true for nonresidential construction. The Council of American Building Officials, an umbrella organization, issues a code for single- and two-family construction that is accepted by the three other organizations.

focus in public policy debate that has waxed and waned over several decades (see Appendix F). Critics of the use of tort litigation and large monetary awards to plaintiffs argue that the risk of such action retards private initiative, is too often unrelated to causal circumstances, and places the party with the greatest capability to pay damage awards—the "deep pockets"— at greatest risk. Others claim that the time lag between the introduction of new technology and the discovery of compensable injury, combined with the high costs of bringing action, leads to underdeterrence of potentially risky new technology. Even those who argue that the U.S. tort liability system, operating optimally, maximizes societal benefits often acknowledge that the system's current operations entail high incidental costs—"friction" losses—that reduce the system's effectiveness. The balance of the impact of tort concerns on innovation, in the private sector in general or on the building industries in particular, remains subject to debate.

GOVERNMENT'S CURRENT ROLE IN FOSTERING NEW TECHNOLOGY

The various government activities described in this chapter, directed at encouraging new building technology, fall into two primary areas:

1. mission-oriented agencies[17] that create a market for new technology by using direct purchase of products and services and R&D funding to seek improved and cost-effective performance that new technology may offer; and
2. agencies as promulgators of policies intended to accomplish broader social goals (e.g., energy efficiency or industrial competitiveness), and to promote development of new technologies that serve these policies.

The committee agreed that both roles—in the context of current federal policies—are appropriate and necessary for all agencies involved in facilities construction and management. Agencies that procure facilities can act broadly to establish an environment conducive to innovation on their building projects. Participation of facility users and agency staff in all phases of project development is important, and these participants can be given responsibility for fostering exploration of new ideas. Even those agencies that do not have direct construction or management responsibility may take an active role in fostering

[17]These are agencies established for specific purposes such as national defense or administering veterans' affairs, compared to those concerned with legislation on more general policy matters in areas such as science and technology or international relations.

new technology in the design or renovation of facilities intended for their own use.

As discussed further in Chapters 3 and 4, responsible government officials must apply their own best professional judgment to ensure that individual health and safety are adequately protected when new technologies are employed. However, just as these officials are expected—as matters of public policy—to meet requirements for open competition and equal opportunity in procurement and environmental protection, so too they should be permitted and encouraged to adopt new technologies that could have broad benefit for the government and the nation. Congressional action or changes in Office of Management and Budget procedures may be needed to facilitate decisions to give priority to new technology.

Government R&D expenditures represent risk money to those agencies for which research and development is not an explicitly assigned responsibility. R&D expenditures by agencies are more frequently the means for trying new ideas rather than developing—in the sense that a private company might—profitable new products. **There appear to be opportunities for improving the contribution of these expenditures to innovation in the construction-related industries overall.**

Although the committee did not undertake a thorough assessment of the productivity of federal building-related research, anecdotal evidence indicates that these programs generally have failed to achieve effective dissemination of research results into practice, as reflected in commercialized products or processes.[18] **The committee found that the criteria government construction agencies use for establishing priorities among areas of potential research should be more clearly linked to the potential value of the new ideas—in practice—that may result from research.** However, agencies lack guidance as to the appropriate balance between costs and anticipated benefits of new technology or the success rates that can reasonably be expected in the field testing of new ideas.

Better guidance can be provided by involving potential commercializers of new technology in all aspects of R&D planning and execution. Expanded programs to encourage solid partnerships between researchers and users of research results are a potentially effective means for enhancing links that turn new ideas into practical innovation.

---

[18]The dissemination problem is apparently quite general. Speaking at a 1989 National Academy of Engineering symposium, industrial innovator Simon Ramo lamented that practically no attempts have been made to educate students in the "art and technique" of turning new ideas into marketable products.

The committee found also that the procedures employed in agricultural, medical, and health care procurement,[19] as well as defense systems, which have been by far the major motivators of government-sponsored R&D effort, highlight a particular need for mechanisms for field testing of new building technology, to ensure that the new technology is likely to perform as promised. **Government agencies, and the building industry in general, lack good means for moving from research efforts to "test-validated" new technology.** The federal laboratories, in particular, could play a more extensive role in testing and prototyping new technology.

It may be possible, under existing regulations and procedures, to use contract incentives, project set-asides, value engineering, and other such programs to encourage innovation. Such specific responses to the committee's findings are discussed further in Chapter 4.

REFERENCES

Building Research Board (BRB). 1988. Building for Tomorrow: Global Enterprise and the U.S. Construction Industry. Washington, D.C.: National Academy Press.

Congressional Research Service. 1991. Transfer of Technology from Publicly Funded Research Institutions to the Private Sector. Subcommittee on Oversight and Investigations of the Committee on Energy and Commerce, U.S. House of Representatives, Washington, D.C.: U.S. Government Printing Office.

Geller, H., J. P. Harris, M. D. Levine, and A. H. Rosenfeld. 1987. The role of federal research and development in advancing energy efficiency: A $50 billion contribution to the U.S. economy. Annual Review of Energy 12:357-397.

General Accounting Office. 1991. Diffusing Innovations; Implementing the Technology Transfer Act of 1986. GAO/TEMD-91-23. Washington, D.C.

Inc. Magazine. 1987. Uncle Sam, research director. (February): 23-25.

Landau, R., and G. N. Hatsopoulos. 1986. Capital formation in the United States and Japan. In The Positive Sum Strategy: Harnessing Technology for Economic Growth, R. Landau and N. Rosenberg, eds. Washington, D.C.: National Academy Press.

MacAuley, P. H. 1990. Federal construction-related expenditures, 1984 to 1991. Construction Review 36(3):iii-xvii.

---

[19]Some examples are food and drug certification and agricultural extension services.

Mokyr, J. 1990. The Lever of Riches: Technological Creativity and Economic Progress. New York: Oxford University Press.

National Science Foundation. 1990. Federal R&D Funding by Budget Function: Fiscal Years 1989-91. Washington, D.C.

Tornatzky, L. G., M. Fleischer, et al. 1990. The Process of Technological Innovation, Lexington, Mass.: Lexington Books.

3

NEW TECHNOLOGY AND INNOVATION IN THE U.S. BUILDING-RELATED INDUSTRIES[20]

As discussed in Chapter 2, government's interests and activities in regard to new building technology can be enhanced to be an even more positive force for innovation. However, the value of government action in this area depends on the overall responsiveness of the U.S. building-related industries to new technology. If these industries are as laggard and resistant to change as some observers have asserted, then efforts to foster new domestic building technology may be unlikely to yield benefits in proportion to their costs. The committee believes that evidence shows otherwise, that investment in the search for new building technology is warranted, and that this search will motivate positive industrial response and broader benefits.

BACKGROUND

As previously noted, new construction (which includes major alterations and renovations of existing facilities) in the United States today is a more than $400 billion per year industry that employs some 6.7 million people. If spending on

[20]A thorough review of the status of U.S. building-related technology and innovation is beyond the scope and resources of this study. Appendix D presents a brief review of recent history, which provided a background for the committee's findings and recommendations.

facility maintenance and operation, as well as materials manufacturing and transport related to construction, is taken into account the building industries as a whole represent a significant percentage of the nation's economy. Construction alone accounts for approximately 7 to 9 percent of gross domestic product (GDP). Building Research Board (BRB) staff estimates that the construction industries combined account for more than 12 percent of GDP. Construction by government at all levels each year typically accounts for less than 20 percent of this total activity. However, if government action to foster innovation in the building industries could increase overall productivity by as little as 1 percent, the likely payoff nationwide could exceed $5 billion annually. Federal programs alone could reap annual benefits of $200 million.

In comparison to the rapid strides being made in electronics, medicine, or biotechnology, some observers feel that innovation in buildings and construction—in the United States, at least—is lagging. The typical user of buildings and construction finds that many of the materials and procedures employed are superficially similar to those of past decades and even past centuries. Some people argue that increasing concerns for environmental and public health implications of building materials and design elements, and progressively diminishing time horizons of decisionmakers focused on immediate financial results, were not conducive to sustained innovation in the U.S. building industries in the 1970s and 1980s. The highly variable and cyclical demand for construction also has made it difficult for builders and suppliers to maintain even medium-term commitments to investment for innovation.

In addition, some observers note that the construction-related industries are dominated, more so than most other industries in the United States, by a male professional and crafts work force. Evidence in other fields suggests that a more balanced and diverse labor force may yield greater numbers of new ideas and different priorities in the pursuit of these new ideas. This raises many questions about labor force recruitment, education, and training that are beyond the scope of the present study. The committee knows of no research that demonstrates gender-specific aspects of innovation.

Nevertheless, the committee observed that the U.S. building and construction industries do work to achieve technological progress, and there are results (see box). Digital controls for heating and cooling systems, power hand tools using lightweight batteries, computer-aided design systems that allow designers and builders to simulate construction of large buildings, roller-compacted concrete, advanced structural design methods, and drywall mounting adhesives are a few recent innovations (see Appendix E).

Some of these innovations originated as inventions emerging from the research laboratory to find widespread application. Fabric and plastic structures, "virtual reality" computer simulations for facilities planning and design, and information management methods that facilitate collection and analysis of facility condition data are new ideas, now being developed, that may become true innovations.

AWARDS FOR INNOVATION

Members of the construction community and their clients recognize that technological innovation often lies at the heart of productivity improvement. The Construction Innovation Forum, a 1987 outgrowth of the Business Roundtable's Construction Industry Cost Effectiveness Project, seeks to foster more innovation through its annual Nova Awards. These awards are given to projects that demonstrate outstanding contributions to quality improvement and cost reduction in the industry.

The first awards, in 1989, recognized the U.S. Postal Service's Kit of Parts, a modular design and construction system for postal facilities, for its improvement of post office function, construction cost, and quality over traditional custom design and construction methods. The engineering firm PBQD's Mount Macdonald Tunnel Ventilation System and the Edward W. Face Company's Face Floor Profile Numbering System (developed to provide designers with an improved means for ensuring that construction measurements are repeatable and accurate) also received awards.

In 1991, Morley Construction Company and the subcontractor, Adams and Smith, were recognized for their development of a means to install seismic isolation devices in the columns of the already constructed Rockwell International Information Systems Center in California. Each column was gripped with friction yokes that supported the building loads during installation, and techniques for sawing through the columns. BE&K Construction Company was recognized for its development of a portable child care center—building and staff—that assures its own and subcontractors' construction workers that their children's care and instruction will be reliable and stable. The center has encouraged employment of women workers and helped to overcome labor shortages.

These cases demonstrate that innovation and new technology can be introduced at many points in the building process and can take varied forms. In addition, besides solving immediate problems and enhancing the industry, these innovations serve broader national goals.

Other groups have begun awards programs as well. For example, the Society of American Military Engineers in 1991 awarded its first Technology Advancement Medal.

However, there are no clear measures or adequate data bases to assess the extent and quality of innovation in the building-related industries. The committee depended primarily on its knowledge of these industries and analyses of innovation in other fields to judge whether the building industries are unique—and underachieving—in their pursuit of new technology and rates of innovation.

LACK OF DATA

The lack of data about the construction-related industries posed particular problems for the committee's work. In the past 10 years, the U.S. Department of Commerce has discontinued or substantially reduced the reporting of some two dozen specific statistical series, such as industry reports on asphalt roofing, clay products, value of state and local construction starts, and economic outlooks for wood products, brick, and sheet metal work.[21] While other data series have been improved, the overall result has been a reduced quality of data on construction-related industries.

In addition, the accuracy of reported data is also in question. For example, reported 1986 spending for improvements to nonresidential buildings was $25.7 billion. A 1986 special Department of Commerce survey (to gather more detailed data) indicated that actual spending may have been $49.4 billion. This major discrepancy, $23.7 billion, represents more than 5 percent of the nation's total reported construction activity and suggests that the uncertainties in overall industry statistics may be substantial.

ATTITUDES TOWARD NEW BUILDING TECHNOLOGIES

A purported special resistance of U.S. building and construction industries to innovation and new technology has long been subject to discussion. A 1970 Department of Commerce report on housing technology described the obstacles to technological change in terms almost identical to another report a decade earlier (Nelkin, 1971). The director of the Department of Housing and Urban Development's notorious Operation Breakthrough program cited the construction industry's structure and general unwillingness to permit research outside of that structure as factors in the failure of federal R&D to achieve substantial payoffs (Finger, 1969).

Studies of innovation in the building industries have, for the most part, been narrowly focused on individual segments of the market (e.g., housing) or specific technologies (e.g., uses of robots), and on the processes of design and construction. Despite the claims that building-related industries are technologically backward, such studies have provided little solid evidence of the problem, and some have shown the opposite result.

---

[21]Reported to the committee by Department of Commerce staff. Also, informal communication with Mr. Kermit Baker, Director of Economics, Cahners Economics.

A recent study of innovation in housing, for example, takes a broader view and suggests that the full extent of innovation has been underrepresented in most previous studies. Further, "a newly constructed residential dwelling is very different from one constructed even fifty years ago, not only in terms of the building components used but also in the techniques employed and the overall performance of the completed structure" (Slaughter, 1991). A survey of literature and interviews with producers identified specific innovations—permanently installed in houses, commercialized in the period 1945 to 1990—that, though only a partial listing, contained 117 specific items. About 80 percent of these items were commercially provided by manufacturers, while the balance were innovations by builders and craftsmen on the job site.

In this experience, residential construction exhibits similarities to the global automobile industry. In this industry, most innovations over the past two decades have involved integrating components and subsystems (e.g., electronic ignitions, digital sensors, and engine control) that reflect innovations in other fields. The resulting automobile of the 1990s is a high-technology product, even though its overall configuration is little changed. A similar evolution is occurring in housing.

Particular needs are a primary motivator for innovation in all fields, and in building this is especially true. Solving problems on specific projects, rather than continuing programs of more generic research, may account for substantial (and largely unrecorded) efforts to develop and adopt new ideas. In addition, the building industry (and particularly residential construction) operates as an assembler of parts and components premanufactured by other companies. These companies operate in areas classified as other industries, and sales to the construction industry often comprise a small percentage of their total markets. Nevertheless, many of these other industries are significant investors in research and development (e.g., chemical manufacturers), and the building and construction industries (and users) reap benefits in new technologies produced as spin-offs of this larger investment.

Most studies of the industry as a whole in the past two decades have been concerned with the decline in productivity observed in macroeconomic statistics, a decline that some people have attributed to failure to maintain a steady rate of innovation.[22] However, little evidence from the construction industry itself is offered to support this attribution, and microeconomic studies have tended to highlight management issues as a primary source of lost productivity on the job site (e.g., Leonard et al., 1988; Smith, 1987; St. Germain, 1985). One major

---

[22]Recent studies (Allen, 1985; Pieper and Allen, 1989) suggest that at least half, and perhaps much more, of the decline is attributable to a shift in the proportions of construction product output—from commercial to residential—and the procedures economists have used to account for inflation.

study characterized U.S. engineering and construction firms as "content to adopt construction technologies pioneered elsewhere," but blamed the design-bid-build strategy typical of U.S. practice—separating responsibilities for design and construction—for weakening incentives to adopt new technologies (Office of Technology Assessment, 1987).

On balance then, the experience of the past several decades fails to demonstrate that the rates of technological advance and innovation in the building industries have been particularly low. The committee concluded that **commonly used norms for assessing innovation rates based on other industries (e.g., aerospace, general manufacturing) are not necessarily an appropriate basis for judging the building industries, and new measures are needed.** However, this conclusion begs the essential question: Are the rates of technologic advance and innovation in the U.S. building industries lower than they could or should be? The committee considered more circumstantial arguments.

ENTRY POINTS FOR BUILDING INNOVATION

The opportunities for adoption of new building technology (i.e., innovation in planning, design, construction, management, or maintenance of facilities) can occur at many points in the process of facility development, through the actions of any of the large numbers of people and organizations involved in that development. The owner, designer, and builder comprise the major participants in that process, but each of these three is in fact a complex group of individuals and organizations that must work together to accomplish the aim of a completed project. Appendix G presents a more detailed portrayal of this complex system.

The key consequence of this complexity, with regard to innovation, is that new ideas and products may enter the process at many different points and will move, in principle, from the lowest levels in the process (i.e., vendors of products, specialist subcontractors, and individual crafts) upward to appear in the final product, the finished facility. These innovations spring primarily from new products, tools, and procedures offered by vendors and from new procedures and relationships initiated by labor, craftsmen, designers, managers, and others working on the project.

Central to the process is the facility owner, who must state his or her needs and employ appropriate resources to meet those needs. The owner's relationship to the facility is, in principle, long term. The facility, viewed as a project, is really not "completed" until the owner replaces, sells and vacates, or otherwise breaks this relationship. When the owner is or represents a large organization, both administration (i.e., related to the organization) and operations (i.e., related to the facility and its occupants) will be considered and will influence what resources are appropriate and how they are employed. Operational aspects are further complicated because the owner often differs from the user. In government facilities, agencies such as the General Services Administration and the U.S. Army Corps of Engineers develop facilities for other governmental units.

Although innovation can start at any point in the facility development process, each of the parties to the process may have particular and differing points of view on the costs and benefits of a proposed new technology (e.g., materials, products, procedures). For example, an improved material may lead to substitution of that material for another, leading in turn to loss of sales for some vendors and loss of work for those who deal with the replaced material. Some participants in the process may oppose new technologies that other participants favor, because the costs and benefits are (or appear to be) distributed disproportionately or simply because the new idea has been proposed by someone else.

New products and processes may then face a tortuous path on the road to becoming innovations. The owner may have great difficulty in determining the ultimate value of the potential innovation and may not even have the opportunity to make that judgment. A sophisticated owner may be able to maintain good information on new developments in each of the fields that represent opportunities for innovation, but most owners must generally depend on designers and builders for information and guidance.

STATUS OF BUILDING RESEARCH

As noted in previous chapters, some observers have expressed concern that the declining position of U.S. construction industries in global markets is due, at least in part, to a declining commitment to research and development in building-related industries. Aggregate spending by industry and government on research and development in these industries—at about 0.4 percent of annual construction output—is well below levels that the industries' aggregate size and importance in the nation's economy warrant (BRB, 1988).

In comparison to other mature industries such as appliances (at 1.4 percent), automobiles (1.7 percent), or textiles (0.8 percent), this spending rate is low. Compared to the construction industry in other countries, the spending is low as well. Estimates assembled in 1983 by the Conseil International du Batiment pour la Recherche l'Etude et la Documentation showed the U.S. rate of building R&D spending at much less than half the rate in Japan, and slightly more than 20 percent of the spending rates in Sweden and Denmark, the nations seemingly most committed to building research (Sebestyen, 1983). Among the leading industrialized nations, only Germany seems to spend at a lower rate, relative to the size of its industry, for building research.[23]

Despite new U.S. programs started since these statistics were assembled (e.g., the Construction Industry Institute, the Civil Engineering Research Foundation, the Corps of Engineers' Construction Productivity Advancement Research program), BRB staff estimates suggest that the current situation is little changed. According to the National Science Foundation, total annual nondefense R&D expenditures in the United States have stayed nearly level at about 1.8 percent of gross national product since 1981 (Jankowski, 1990). In West Germany and Japan the 1988 spending rates were approximately 2.6 and 2.9 percent, respectively, up some 30 to 40 percent over the past decade.

As has already been noted, research does not necessarily lead to innovation but is more likely to do so when close ties exist between R&D and the potential users of the resulting new technology. The committee is unaware of any comprehensive analysis or data source that would enable analysis of the strength of this relationship and the factors that influence it in the building-related industries. Nevertheless, there is evidence that supports its importance.

A 1990 evaluation of Japan's construction industry, sponsored by the National Science Foundation, found that aggressive and highly productive research spending by industry and government have placed that nation at the forefront of construction technology. Despite substantial dependence on ideas initially developed in the United States and other countries, Japanese industry leads the United States in many areas and is gaining rapidly in virtually all areas examined. The quality of new facilities in Japan equals and often exceeds that of new construction in the United States, and Japanese industry's efforts may lead to "new breakthrough technologies" (Tucker et al., 1991).

A subsequent reconnaissance of major Japanese construction R&D facilities, sponsored by the Civil Engineering Research Foundation (1991), attributed much of Japan's apparently substantial ability to move research results into practice to the very close ties between researchers and design and construction professionals. Major construction companies (which in Japan are frequently responsible for design as well) maintain their own well-funded research facilities and research programs that extend beyond these facilities to the project site. Many of these companies rotate professional staff or otherwise ensure that researchers experience field conditions and that field professionals participate in research. Government research facilities may be used jointly by government and private sector researchers, and government uses specific projects to ensure a "market"

---

[23]However, such data on the building industries in most countries are difficult to assemble and notably less reliable than manufacturing industry statistics (Brochner and Grandinson, 1991).

(i.e., a means for recovering research investment costs) for research in high-priority areas.

The importance of the market linkage is demonstrated in the United States as well. For example, Du Pont and other chemical companies have in recent years used their research to develop a number of polymeric solid interior surface materials for countertops and other work surfaces. These materials, with such brand names as Corian, Avonite, and Nevamar, offer easy workability, high durability, and a wide range of colors and visual textures that have led to their increasingly widespread use in hospitals, restaurants, and residential kitchen and bath applications.

DETERRENTS TO TRANSFER OF NEW TECHNOLOGY

As such cases illustrate, once new ideas are produced—in the research laboratory, on the job site, or elsewhere—they must be transferred into general practice to become innovations. The structure of the building industries, (in which many smaller firms operate in narrow geographic areas and lack vertical integration), a complex regulatory framework with many locally administered building codes, concentrated attention on reducing the initial cost of facilities (often at the expense of higher operating and maintenance costs), and exposure to litigation that increases the business risks of new products and processes—all are among the factors that have deterred the spread of new ideas.

Materials and equipment manufacturers and their sales representatives play a role as sources of new technology and innovation, but this role has generally been underestimated in studies of innovation in the building-related industries. An earlier study by the BRB (1988) attempted to include this source and concluded that roughly two-thirds of the annual spending for U.S. building-related research comes from manufacturers, primarily for new product development and marketing support activities. Solid surface materials (which require little or no postfabrication finishing) are only one of many innovations that have entered the building-related industries from this source (refer to Appendix E).

Of the remaining annual R&D spending, BRB staff estimate that roughly two-thirds of the balance (i.e., about 22 percent of total spending overall) supports activities in university research facilities. The separation of researchers from the construction contractors and facilities managers who comprise the potential users of research results (in direct contrast to the Japanese situation described previously) deters transfer of new ideas into practice.

Owners may choose among a variety of forms for the contractual relationships with designers and builders, and some of these relationships are more congenial to innovation. Award of contracts based on a lowest-bid procurement strategy, for example, increases the bidder's risk in trying new ideas because the bidder must bear the costs while the owner may benefit over the longer term. Open bidding that encourages competition makes it more difficult to screen unqualified suppliers or give preference to those inclined toward innovative practices. Rigid budgeting and construction contracting practices that focus on low price often preclude higher spending to procure a facility that will cost less to maintain or will yield more reliable service in the future, although innovations that reduce initial cost may be encouraged. Preselection of bidders, design-build procurements, and other mechanisms can be used to encourage cooperation among the participants in the process and thereby ease the introduction of new technology (Lemer, 1991).

Poor communication of advances in knowledge in general may be another important deterrent to the transfer of technology and subsequent innovation in the building-related industries. Testimony of agencies and researchers suggests that many potentially useful new ideas fail to reach a broad audience of potential users. While informal networks of communication among researchers and users are effective, dependence on the National Technical Information Service (NTIS)[24] and inadequate programs for personnel exchange (particularly with overseas research institutions) have, in the building industries, failed to provide the information exchange needed for effective innovation. In addition, there is little transfer of researchers among building research and other fields, which limits the cross-fertilization that has been seen to occur at many leading industrial research laboratories.

The Construction Industry Institute (CII), based at the University of Texas, was established to motivate a closer partnership among academic, government, and private sector members of the industry and thereby improve communication as well as commonality of direction in solving industry problems. The CII uses a portion of the funds provided by its member participants to sponsor researchers in studies of current problems. An emphasis in this research has been placed on management-related issues. The Civil Engineering Research Foundation (CERF) and the National Institute of Building Sciences (NIBS), both based in Washington, D.C., also conduct or sponsor research in similar topic areas. All of these organizations have encountered difficulties achieving broad dissemination and widespread adoption of their work.[25]

Centralized organizations such as the CII, CERF, and NIBS are most accessible to larger firms but are less effective in reaching the many medium-

---

[24]The NTIS distributes government publications and many other documents to the general public.

[25]The BRB has encountered these problems as well. Inability to realize commercial advantage or otherwise recover costs is a serious deterrent to technology transfer activity.

and small-sized firms operating in the building-related industries. The very widely distributed nature of these industries inevitably slows the spread of new ideas.[26]

Some corporate and government owners have improved integration in the delivery process by forming design and development teams that bring together corporate staff and the construction contractor, along with the designers and sometimes major building subsystem suppliers. These teams come together in a variety of ways, and such examples as CIGNA Corporation's headquarters buildings, the operations of the major Japanese general contractors, and some design-build procurements are practical illustrations of team formation and management (Lemer, 1991).

The committee concluded that the current structure of the building-related industries poses inherent problems for transfer of new ideas to practice. These problems probably make the building industries less susceptible than other sectors to applications of new technology. Efforts to provide a better integration of the process from the development of new ideas to their application in practice could enhance innovation rates in the building-related industries.

LIMITED OPPORTUNITIES FOR INNOVATION

Buildings and other constructed facilities typically have long service lifetimes of 30 to 40 years. Many structures survive much longer, sometimes valued more for their historic associations than their functional potential. Government buildings often have projected lifetimes of 100 years, although renovations may be required during this long life. Most owners and users of these facilities have a much shorter perspective and get only one or two chances over the course of a career or lifetime to participate actively in the building process. Thus the marketplace of users and owners, identified in most fields as a driving force for innovation, is severely constrained in the area of building.

In many fields of manufacturing, product cycles of three to five years permit innovators to recover research and development costs at acceptably rapid rates. Computers are currently an extreme case: one leading maker of work-stations popular with the developers and users of computer-aided design has introduced eight new generations of computers in less than 10 years (Bulkeley,

---

[26]In Sweden, with strong government support for the building industry and a much smaller set of participants, ideas spread more quickly. This situation sometimes causes problems when new ideas receive limited testing before widespread application (H. Westling, Royal Institute of Technology, Stockholm, personal communication with A. Lemer 1991).

1991). To the extent that the technology in question is used in applications with shorter service lives (e.g., microcomputers used by designers and facilities managers, telecommunications or HVAC[27] controls fitted to modular networks, paints and other interior finishes, just-in-time delivery scheduling by contractors),[28] higher rates of innovation can more reasonably be expected than in most aspects of buildings.

Commercial developers, in particular, are poorly suited to drive many aspects of building innovation because they typically must seek to recover the costs of the project quickly, and tenants may be hesitant to pay higher rents (which in turn lead to higher capital value and sales price for the developer) for innovation that does not clearly yield direct benefits. Major corporations or other large institutions—or governments—that build for their own long-term use (and are thus able to reap the advantages of new technology that enhances a building's lifetime performance) are more likely to find value in the search for building innovation.

Major constructors stand to benefit directly from improvements in construction speed, safety, and reliability, and thus may be prone to search for innovation in the construction process. Owners benefit as well, to the extent that savings are passed along or the resulting facility's quality is increased. Architects and engineers responsible for facilities planning and design may also seek innovation that improves their own work, and they are generally responsive to new products and processes that offer their clients improved performance at an affordable and competitive cost. However, buildings and other constructed facilities are subject to a large number of prescriptive controls placed on facility design, construction, and operation to ensure the safety and health of building occupants and neighbors.[29] These controls, embodied in standards and guide specifications, building codes, and procurement regulations, contribute to social well-being but inevitably constrain the individual ability to innovate, both by preventing precipitous introduction of untried products and procedures and by

---

[27]Heating, Ventilating and Air Conditioning.

[28]The "lifetime" of a typical building construction project is 18 to 36 months.

[29]Performance specifications state the results required rather than specific materials and dimensions selected to deliver these results. Such specifications, now widely used for structural systems and slowly appearing in other aspects of facilities, are purported to reduce one obstacle to innovation. Designers and owners, particularly government agencies that must allow open, competitive bidding, are sometimes reluctant to adopt performance specifications, fearing that available forms for such specifications cannot adequately define the requirements.

imposing costs on those who seek to have these products and procedures accepted.

Concern for consumer protection introduces similar problems in other areas, most notably the food and drug industries. The government's Food and Drug Administration must approve new drugs and food additives for human consumption and, by establishing the standard of judgment for safety, to some extent moderates the burden of risk assumed by private sector innovators in these industries. However, many observers agree the system is very costly. Examples representing less risk to human health and less cost include the Underwriters Laboratory, Inc. (UL), which tests and certifies a wide range of electrical products.

Except in limited areas such as UL, the building-related industries have no comparable government-endorsed mechanism for approving new technology. The professional societies and model building code organizations that review new technology can recommend its acceptance, but it is then up to each government agency, state, or local government to accept new products or procedures on their projects or within their jurisdictions. Independent evaluations of demonstrations of new technologies in transportation facility construction[30] have reportedly helped speed adoption of these technologies.

Adapting practices currently used in Japan (CERF, 1991) and Europe, CERF has proposed that government and industry should join to establish one or more Innovation Test and Evaluation Centers (ITECs) to test, demonstrate, evaluate, and document innovative building technology (Bernstein, 1991). These ITECs would help reduce the perceived risk of applying new technology by assessing more thoroughly the likely actual risk.

Such centers might then help the industries to deal with the seemingly progressive growth in the public's aversion to risk in general, particularly to the types of long-lasting risks to public health and safety that are commonly encountered in the building-related industries.[31] Designers, owners, and managers are understandably reluctant to try new technology that may lead to expensive litigation if an accident occurs or the technology fails to perform adequately. Consider asbestos: Millions of dollars are being spent to remove this once popular fireproofing and insulating material that is now seen to pose an unacceptably high risk of cancer to building occupants. This example, admittedly extreme, nevertheless influences many people who might otherwise be inclined to bring to market a product or other innovation with an uncertain level of risk.

---

[30]Funded by the U.S. Department of Transportation.

[31]A more complete discussion of risk in and around buildings may be found in the BRB report Uses of Risk Analysis to Achieve Balanced Safety in Building Design and Operations (McDowell and Lemer, 1991).

On the other hand, the system for dealing with tort liability is designed to protect valid public as well as private interests, and some committee members concluded that this system is generally effective—albeit costly—in responding to the occasional losses associated with the unavoidable risks of new technology (see Appendix F). **Improvements in the efficiency of the tort liability system (i.e., the relationship of the costs of the system to the actual losses being addressed) are necessary to reduce the system's apparent deterrent effect on innovation.**

In addition, insurance has proved to be an effective means of distributing risk to better match the anticipated rewards of new technology. In government construction, agencies are, in principle, self-insuring. However, failure to budget for correcting losses associated with trying new ideas means that the government is, in effect, uninsured. The threat of financial loss and damage to professional careers therefore seriously limit opportunities for innovation.

PLACE OF UNIVERSITIES AND OTHER EDUCATIONAL INSTITUTIONS

Educational institutions play an important role in training the architects, engineers, and other building professionals who produce, perfect, or accept and apply much of the new building technology that enters practice. Budgetary pressures that threaten the levels of education and the production of academic researchers have been a subject of frequent concern among academics (e.g., National Research Council, 1985).

Some practicing professionals argue that university courses in architecture and several areas of engineering are poorly tailored to fit the needs of professionals in the building industries. They assert that newly graduated architects and engineers have little understanding of the practical implications of theory and almost no familiarity with current practices in design and construction. These young professionals thus lack an essential ability to deal in practical terms with the cross-disciplinary judgments that are inherent to facilities design and construction. In addition, they typically have little exposure to social science, organizational, and management factors that influence implementation and long-term performance of facilities.

Some educators counter this argument by observing that practitioners need a firm grounding in basic principles to enable them to keep pace with rapidly changing technology. These educators find it increasingly difficult to deliver what they would view as a well-trained professional graduate within the constraints of the four-year college program still typical for engineering. Graduate programs are already the predominant source of first professional degrees in architecture and landscape architecture, and the master's degree is increasingly essential for engineers. Throughout the building-related industries, apprenticeship and other on-the-job training are recognized as essential elements of eduction.

In general, education can create an intellectual environment more or less conducive to innovation. Committee members noted that levels of training and education among European craftsmen, design professionals, and construction companies appear to be generally higher than those in the United States and attribute higher European rates of new technology development, at least in part, to this factor.

REFERENCES

Allen, S. G. 1985. Why construction industry productivity is declining. Review of Economics and Statistics 67(4):661-669.

Bernstein, H. M. 1991. Is innovative engineering becoming extinct in the U.S.? American Consulting Engineer 2(4):37-38.

Brochner, J., and B. Grandinson. 1992. R&D cooperation by Swedish contractors. Journal of Construction Engineering and Management 118(Mar):3-16.

Building Research Board (BRB). 1988. Building for Tomorrow: Global Enterprise and the U.S. Construction Industry. Washington, D.C.: National Academy Press.

Bulkeley, W. M. 1991. Computers become a kind of commodity, to the dismay of makers. Wall Street Journal CCXVIII(47):A1.

Civil Engineering Research Foundation and Japan International Research Task Force. 1991. Transferring Research into Practice: Lessons From Japan's Construction Industry. Washington, D.C.

Finger, H. 1969. The goal of Operation Breakthrough. Journal of Housing (December):587.

Jankowski, J. E. Jr., 1990. National Patterns of R&D Resources: 1990 Report number NSF 90-316. National Science Foundation, Washington, D.C.

Lemer, A. 1991. Team-building and quality buildings. Design Management Journal 2(2):54-58.

Leonard, C. A., P. Fazio, and O. Moselhi. 1988. Construction productivity: Major causes of impact. AACE Transactions D.10.1- D.10.7.

McDowell, B., and A. Lemer, eds. 1991. Uses of Risk Analysis to Achieve Balanced Safety in Building Design and Operations. Building Research Board, National Research Council. Washington, D.C.: National Academy Press.

National Research Council 1985. Engineering Graduate Education and Research, Washington, D.C.: National Academy Press.

Nelkin, D. 1971. The Politics of Housing Innovation. Ithaca: Cornell University Press.

Office of Technology Assessment 1987. International Competition in Services: Banking Building Software Know-How. Washington, D.C.: U.S. Government Printing Office.

Pieper, P., and S.G. Allen. 1989. Why construction industry productivity is declining. Review of Economics and Statistics 7(3):543-549.

Sebestyen, Gy. 1983. Measurement, Accounting, and Taxation for Research and Development: Volume and Support of Construction R&D. Study submitted to the Board, Conseil International du Batiment pour la Recherche l'Etude et la Documentation (CIB), Rotterdam.

Slaughter, S. E. 1991. "Rapid" innovation and the integration of components: A comparison of user and manufacturer innovations through a study of the residential construction industry. Ph.D. thesis submitted in the field of Management of Technology, Massachusetts Institute of Technology, Cambridge, Mass.

Smith, A. G. 1987. Increasing onsite production. AACE Transactions K.4.1-K.4.14.

St. Germain, E. F. 1985. Effects of Environmental Factors on Worker Productivity in the Construction Industry. Report No. GRAI8610, Washington University, Seattle.

Tucker, R. L., et al. 1991. JTEC Panel Report on Construction Technologies in Japan, Final Report, Japanese Technology Evaluation Center, Loyola College, Baltimore, MD.

4

WHAT SHOULD THE ROLE OF THE FEDERAL AGENCIES BE IN FOSTERING NEW BUILDING TECHNOLOGY?

Against this backdrop of the U.S. building-related industries as a whole, the committee considered the question, What should the role of the federal agencies be in fostering new building technology? As a purchaser and user of facilities, as well as a funder and performer of research on building technology, the federal government already plays a substantial role in the field. Should this role change?

Opposing trends make it difficult to foresee the place of the building-related industries in the future U.S. economy. Some observers suggest that declining defense expenditures may spur growth in government spending on domestic public works and a shifting of production resources into this field. Others note that U.S. population growth has slowed and the nation's people are aging, which suggests that fewer homes and offices (and consequently less public works infrastructure) will be needed in the future. Some of these latter observers feel that the widespread 1990 U.S. real estate recession is only an initial demonstration of the market's response to gross overbuilding in many regions.

The committee noted, however, that regardless of aggregate trends, the geographic distribution of the nation's population is likely to continue shifting. The substantial capital stock of public facilities will continue to provide vital support for the private sector's activities and will need maintenance and periodic refurbishment. Technological advances and rising expectations will continue to warrant the upgrading and replacement of obsolete facilities. Even if their share of the U.S. economy continues to decline as other sectors experience more rapid growth, the building-related industries will continue to make important contributions to the quality of life.

Some observers suggest that the U.S. industry could be on the brink of major change (Office of Technology Assessment, 1988). This industry will either undergo basic shifts in structure and production processes or lose out to foreign competition. The committee agrees. The technological gap between nations with higher and lower per capita incomes is narrowing in many fields, including much of the building-related industries. Declining relative costs of transportation and communication are likely to continue the trend toward multinational production of goods and services. Business for U.S. construction-related industries will not get any easier.

The committee noted that we lack adequate statistics to form a clear and comprehensive historical picture of the development of new technology and innovation in the building-related industries. Nevertheless, on the basis of its review of evidence and the structure of these industries in comparison to others, the committee concluded that for at least the past two decades, the rates of U.S. development of new building technology and innovation have been lower and the role of government has been less positive than they should be, in terms of both the industry's potential capabilities and the nation's welfare. On all counts—**to achieve better cost, quality, and performance in its own facilities; to enhance the quality of life in the United States generally; and to enhance U.S. industry's productivity in international markets—government should seek to foster new building technology and innovation.**

INSTITUTIONAL PERSPECTIVE

In contrast to many other countries, there is no single U.S. government agency with explicit responsibility for representing or encouraging enhancement of the nation's construction industry as a whole. The Department of Commerce and the Office of Science and Technology Policy share executive branch concern for the nation's technology and industrial competitiveness, but seldom address issues of construction and facilities.[32] In addition, there is little basis for communication among the agencies that undertake construction or manage facilities as accessories to their primary missions and the policy-oriented

[32]A major exception is the focus on the construction industries of Japan and the United States in trade negotiations of the late 1980s and early 1990s.

agencies that may include construction, building products and equipment, or facilities themselves (i.e., housing or highways) within their broader purview.[33]

Institutional analogies have from time to time been drawn between the U.S. farm and construction industries, both characterized by many small producers spread across the country, and proposals have been made that there should be a construction equivalent to the Department of Agriculture or a U.S. government equivalent of other nations' ministries of construction. In a historic example, the national crisis of the Great Depression of the 1930s—during which the total annual rate of construction in the United States dropped to one-third of its average in the late 1920s—fostered creation of the Public Works Administration (PWA).

However, the diversity of interests among federal construction agencies and the many organizations active in the private sector have made the construction sector as a whole generally unresponsive or antagonistic to such proposals. Although the PWA survived for some years, its role as builder was largely relinquished to local government or supplanted by special-purpose agencies (Craig, 1984). Other centralized building programs of the era, such as the Works Projects (originally Progress) Administration, were dismantled as the nation went to war in the 1940s.

Analogies may also be drawn between the U.S. construction and aerospace or shipbuilding industries. In both of the latter, the purchaser/owner customarily takes delivery of a fully finished aircraft or ship from a single responsible contractor, rather than entering into separate agreements with providers of design, construction, and interior furnishings, fixtures, and equipment.

Committee members noted that U.S. shipbuilding 20 to 30 years ago offered many parallels to construction today. In the late 1970s and early 1980s, shipbuilding underwent a "quiet, relatively unpublicized transformation" marked by increasing labor productivity and improved product quality (Marine Board, 1984). The transformation was spurred by loss of competitive commercial shipbuilding awards to foreign shipyards and subsequent actions within the industry to learn why. One major action was the 1976 formation, jointly by government and industry, of the National Shipbuilding Research Program to make Japanese technological advances accessible to U.S. shipbuilders. Some U.S. shipyards had Japanese personnel working side by side with U.S. counterparts to shift U.S. production practices. However, despite some bright spots, the U.S.

---

[33]Again, there are exceptions. The U.S. Army Corps of Engineers, which undertakes a great deal of construction, and the Department of Transportation, primarily a policy-oriented agency, are jointly responsible for exploring magnetic levitation (maglev) technology for high-speed ground transportation.

industry as a whole has failed to adjust well to global competition and declining military procurements, and has continued to shrink.

The committee also considered more recent centralized government R&D programs, such as the Defense Advanced Research Projects Agency, widely credited with sponsoring major technological advances in electronics and guidance systems, but it concluded that such institutions would match poorly with the structure of the industry. Hence, in considering what the precise nature of government's roles in fostering new building technology might be, **the committee assumed as a starting point that construction responsibilities will remain distributed among the several agencies now holding these responsibilities. Establishing a new construction agency or centralized construction program, or any substantial consolidation of policy responsibility for the building-related industries, is unlikely to be achievable or appropriate.** Such options were given no further consideration.

ALTERNATIVE ROLES FOR GOVERNMENT

On the basis of its findings described in Chapters 2 and 3 and the assumption that current allocations of agency responsibilities will remain for the most part unchanged, the committee considered the range of strategic roles government might play in fostering new building technology. This range was characterized in terms of three broad options representing increasingly proactive levels of government involvement in the U.S. building sector's innovation processes:

1. **Business as usual** would represent a continuation of current policies and levels of funding. This is the baseline against which any recommendations for change must be compared. Current events suggest that the proportions of federal agency construction budgets may, in the future, shift toward civilian facilities and decline in absolute terms for at least the next decade. State and local governments could, in aggregate, experience substantial growth in construction for renewal of physical infrastructure, renovation and expansion of educational facilities, and accommodation of increased work loads associated with programs formerly administered at federal levels. Steady federal commitment to building-related R&D is assumed because benefits would accrue to more active state and local governments as well as federal agencies.
2. **Active mission-oriented pursuit** of new technology would involve agencies taking definite action to encourage the development of new building technology that can be applied in their own facilities projects. General policy actions might motivate the agencies, for example, (1) required "new technology set-asides" in procurement to permit sole-source procurement of new technology and to fund testing and demonstration; (2) required annual progress reports by

each agency to the Office of Science and Technology Policy; or (3) establishment of presidential awards for new building technology demonstrations. Mechanisms would be developed to overcome fragmentation and encourage common direction among agencies and within the industries. A new commission or institutional focus within an existing agency could be such mechanism. Individual agencies could respond by assigning responsibility and resources to a centralized "building innovation office" or contracting with one of the national laboratories for assistance in identifying promising projects and technologies. Individuals and firms involved in developing and demonstrating new technologies with commercial potential would be given patent and copyright protection to encourage the dissemination of successful new ideas. The return on federal investment would be realized in direct benefits to the agency applying the new technology, and the expenses of inevitable uncertainties and occasional failures would be accepted by the executive branch and Congress as the cost of the program. Establishing a system for evaluating and reporting the results of new technology applications would assume that this return on investment is recognized and favorable.

3. **Enhanced federal research, development, and demonstration support** activities would be coordinated and possibly expanded (or new ones created). Effort would be made to draw more effectively on state and local governments and the private sector. A national program, based perhaps on the model of the National Science Foundation or the Strategic Highway Research Program, both congressionally funded, or the Army's Construction Productivity Advancement Research program, would provide focus and leadership. A coordinated effort could be made to optimize return—from a national perspective—on the R&D and demonstration investment of mission-oriented construction agencies. Tax disincentives that discourage private investment in new building-related research and development could be reduced and actions taken to decrease the costs of the tort liability system, limit the diversity and restriction of building regulations, and generally lessen the regulatory and administrative impediments to the application of new technology (e.g., restrictive codes and guide specifications), consistent with the protection of safety, health, and environmental integrity.

The problems of public policy toward technological innovation are exceedingly complex, as illustrated by several potential paradoxes (David, 1986). First, efforts to speed the rate of innovation in industries supplying capital goods may create expectations of more rapid obsolescence for those users who consider adopting new technology when it first appears, thus encouraging a "wait-and-see" attitude that actually delays broad adoption.[34] Second, tax and other subsidies for research and development can reduce the cost of initiating new ideas and thus diminish the value of being first to develop or adopt new technology, when there are many competitors in the same business. Finally, delaying introduction of technical standards to facilitate continuing R&D investment can retard effective application of technologies whose benefits depend on compatibility and system integration.

Through such paradoxical effects, policy initiatives may have exactly the opposite of their intended impact. The case for encouraging wider adoption of new technologies in any particular area must therefore be considered primarily within that particular context. Policies and programs that will encourage the development of new construction materials, for example, could be totally ineffective in fostering innovation in building electrical systems. The committee determined that a "top-down" consideration of broad strategic directions would be a useful first step in formulating recommended government action to foster new building technology, but only a first step. Specific recommendations for action would have to be considered in the context of the agencies called on to act.

EVALUATION OF STRATEGY OPTIONS

As noted in Chapter 2, the committee identified at least three key reasons why government, in general, and the federal government in particular, might seek new technology and innovation in building, and determined that the federal government should, on all three counts, seek to foster new building, technology and innovation. These reasons become then the objective to be achieved by government strategy to foster new technology. The committee considered strategy proposals in terms of their likely contribution to these three objectives:

1. better cost (initial or over the course of a building's life cycle), quality, and performance in government facilities;
2. enhanced quality of life in the United States generally (and world-wide), through encouraging better cost—initial or life cycle—quality, and performance in private sector building; and
3. enhanced U.S. industrial competitiveness in international markets.

---

[34]Problems such as building obsolescence and design actions that can be taken to avoid these problems are the subject of another Federal Construction Council-sponsored Building Research Board study.

The private and social costs of achieving these objectives must be both affordable and in reasonable proportion to the benefits of achievement.

Table 4-1 summarizes the results of the committee's discussion of the range of strategy options in terms of the increasingly proactive levels of government involvement. The committee agreed that "business as usual" is likely to lead to increasing penetration of foreign building products, equipment, and technology into the U.S. market, with resulting losses of employment and output for U.S. companies. To the extent that displaced sales and jobs are replaced in other sectors of the economy, and as long as conditions of international trade make it possible for U.S. builders and consumers to purchase foreign technology without paying significant penalties, the negative consequences of this strategy may be limited. However, experience in manufacturing suggests that the "new" jobs created are lower skilled and lower paid. Moreover, government agencies may suffer particular hardship if restrictions on the purchase of foreign goods and services limit the agencies' ability to obtain performance and costs available in the private sector.

Agencies may gain some benefits by taking action to encourage the development of new technology to be applied in government building programs. Such action might be similar to current support for technology advancement in areas with defense applications. However, as has sometimes been the case in defense, foreign holders of advanced technology may be the only reasonable recipients of such support and the primary beneficiaries of consequent innovation.[35] In addition, broader public benefit will be realized only to the extent that effective transfer of new technology to the private sector is accomplished.

Increases in federal support for new building technology development in general would have to be substantial to make a significant impact on innovation rates overall. In the current political climate of sizable government deficits at all levels and federal disengagement from activities that can be distributed to state or local levels, major growth in funding for building-related R&D or demonstration projects seems unlikely. However, modest increases of 15 to 25 percent may be possible, and targeting of particular technologies could increase the likely return on this investment. Research centers such as those established to focus on cement technology and large-scale structures are one proven method for accomplishing this targeting effectively.

---

[35]For example, Japan's Hoya Corporation received $8 million in U.S. research funds for the development of high-purity glasses for use in large lasers. The techniques they are perfecting make the company a strong competitor in the markets for television and photolithography lenses (the latter are used for microchip manufacture) (Eisenstadt, 1991).

Table 4-1 Evaluation of Strategy for Fostering New Building Technology

Levels of Involvement	Anticipated Achievement of Objectives		
	Better Cost, Quality, and Performance in Government Facilities	Enhanced Quality of Life	Enhanced U.S. Industrial Competitiveness
Business as usual: Baseline; continuation of current policies and levels of funding	More rapid foreign technology growth may put government at a disadvantage because of "Buy America" programs' raising costs or reducing performance compared to today's levels	Increased introduction of foreign technology could maintain overall U.S. quality of life	U.S. building industries likely to be under increasing pressure of foreign competition
	Achievement Relative to Base Case		
Active mission-oriented technology pursuit: Agencies take definite action to encourage new building technology applied in their own projects	Short-term program costs may increase to cover testing, evaluation, and less successful demonstrations; public benefit will increase to the extent that successful technology is transferred	Likely to have limited effect on industry innovation rates; hence little impact overall	Likely to have limited effect, unless programs are restricted to U.S. firms; effect likely to be restricted to major products and project types for which U.S. firms compete overseas, such as transportation facilities, construction management, and equipment sales
Increased federal research, development, and demonstration support: Expanded or new funding of building-related research and development activities	Substantial increase in funding required to have significant impact; targeting of specific technologies may increase likely payoff	Could increase innovation rates in targeted technologies	To the extent that domestic innovation rates are increased for exportable technologies, competitive position of U.S. firms will be enhanced

Because of government agencies' leadership role in the construction sector and because agency programs are a primary means for effecting government policy, the committee agreed that government-wide "business as usual" is not an acceptable option. **Federal agencies have a definite responsibility and role in fostering new building technology and should take action to do so.** However, variations among agencies' missions and resources preclude adoption of a single prototype strategy for all agencies.

The committee concluded that **each agency must tailor activities to foster innovation to the specific characteristics of its programs.** Each agency may then, in principle, continue with business as usual on many projects, while at the same time pursuing new technology in areas likely to yield greatest benefit to that specific agency and others sponsoring research and technology transfer efforts. In addition, even business as usual can be carried out in ways that encourage agency staff, designers, constructors, and suppliers to demonstrate new technology. The committee's recommendations for government action, presented in Chapter 5, thus include three elements of strategy at increasing levels of proactive involvement:

1. Federal agencies responsible for developing and operating facilities should seek to encourage innovation in their own projects.
2. These agencies, with the assistance of broader government programs that span the gulf between agencies, should undertake broader activities to enhance the effective transfer of technology from government applications to the private sector.
3. Agencies should work together and separately to support targeted efforts to develop and transfer new technology in specific high-priority areas.

Government's role in fostering new building technology will depend on the participation of a variety of agencies. Effective coordination may be needed in some areas, and monitoring of progress will be important overall, to ensure that government resources are well utilized. Chapter 5 discusses the committee's recommendations for implementing this strategy to enhance building innovation.

REFERENCES

Craig, L. 1984. The Federal Presence: Architecture, Politics, and National Design. Cambridge, Mass: MIT Press.

David, P. A. 1986. Technology diffusion, public policy, and industrial competitiveness. In The Positive Sum Strategy: Harnessing Technology for Economic Growth, Landau, R., and N. Rosenberg, eds. Washington, D.C.: National Academy Press.

Eisenstadt, G. 1991. Watch out, Corning. Forbes (August 19): 64-68.
Marine Board, National Research Council. 1984. Toward More Productive Naval Shipbuilding. Washington, D.C.: National Academy Press.
Office of Technology Assessment. 1988. Technology and the American Economic Transition: Choices for the Future. Washington, D.C.: U.S. Government Printing Office.

5

IMPLEMENTING AN EFFECTIVE ROLE

The three-part strategy—encompassing adoption of new technology for agency projects, enhanced technology transfer, and targeted support for new technology development—responds to the government's overall responsibility to take a leadership role in fostering new building technology. However, the committee recognized that agencies differ in their missions and resources, so this role must be tailored to specific agencies, nongovernmental organizations, and individuals. Each of these agencies, organizations, and individuals (as well as the nation as a whole) will benefit from enhanced innovation in the building-related industries, but there are risks involved also. In any particular instance, a new technology may not work as expected. In turning their attention to the specific actions needed to implement an effective strategic role for government in fostering new building technology, committee members considered carefully how to balance individual and aggregate risk and reward, from the points of view of agencies' programs, the U.S. building-related industries, and the nation as a whole. The committee's recommendations seek to achieve this balance.

RESPONSIBILITIES FOR TAKING ACTION

Successful development of the government role in fostering new building technology will require action by several groups:

- Mission-oriented government agencies, both those responsible for facilities development and those responsible for policy influencing the building-related industries, are called on to actively encourage the development and broad application of new building technology.
- Government oversight agencies and groups (e.g., the Office of Management and Budget, committees of Congress) must provide a supportive enabling environment for innovation.
- Government agencies and other organizations responsible for funding and conduct of research and development are called on to pursue more aggressively the broad practical application of new technologies that offer significantly enhanced productivity and performance in the building-related industries.
- Private enterprise providing goods and services to government must continue to pursue new technology and work in partnership with government to devise more effective ways to reduce and mitigate the technical and commercial risks that deter innovation.

In addition, all of these groups must work to enhance public understanding of the value of new building technology to our quality of life. Better understanding will lead to more informed decisions about new technology and ultimately to enhanced innovation. An institutional focus within the federal government is needed to provide strong leadership.

INSTITUTIONAL FOCUS NEEDED

As discussed in Chapters 1 and 2, innovation occurs when new ideas are put into practice. The three-part strategic role envisioned for government is meant to foster both the generation of new ideas and the application of these ideas in the design, construction, and management of constructed facilities. However, the degree to which the second step is accomplished, (i.e., putting new ideas into practice) is the measure by which the success of government's role in fostering new building technology should be judged.

An institutional mechanism is needed to focus attention on technology in the building-related industries, to exercise leadership in implementing strategy to foster new building technology, and to monitor and report on the progress of government's efforts. The lack of such leadership and effective information to support technological assessment of the building-related industries places this important sector of the nation's economy at a serious disadvantage in public policy forums.

As explained in earlier chapters, the industry is composed for the most part of many small and regionally constrained firms that construct facilities and produce building products and materials. There is no single agency or program in the federal government with comprehensive responsibility for dealing with issues of construction and facilities. At the same time, there are many agencies that build facilities and can benefit directly from innovation in these facilities. These agencies have separate missions, operate independently, and are in some senses competitive in their traditions and administrative procedures. For these reasons, the committee concluded that an institutional focus is needed, within or closely linked to government, but outside of any single agency's existing facility programs. Responsibility and resources for coordinating government building-related innovation strategy, taking positive leadership in implementing this strategy, fostering action by construction agencies, and evaluating progress should be assigned to the office or agency in which this focus is established.

This leadership organization could be placed within a government agency (e.g., a unit of the Department of Commerce or the Department of Housing and Urban Development), a federal laboratory, or a government-related but independent organization (e.g., the National Institute of Building Sciences), but it should be clearly separated from the development of facilities or spending for building-related research. Models for the creation of a new organization exist as well, such as the Council on Environmental Quality, and should be given consideration. In either case, a regular and broadly distributed strategy statement and progress evaluation report, prepared perhaps biennially, would facilitate industry involvement in the direction of effort to develop and disseminate new building technology.

This leadership organization or office should be given responsibility and authority to support technology development and demonstration in government facilities. The Department of Transportation, for example, has implemented several such demonstration programs in the past, using federally supported state and local transportation projects to demonstrate new products or procedures.

The organization or office would provide funds for a relatively small number of projects each year, probably 10 or fewer. Funds would come from a 2 to 5 percent set-aside or "tax" on all agencies' construction appropriations, similar in form to the mechanism now used to fund art in public places. Agencies would then effectively compete for support of their specific projects.

Such a funding program should be established with a limited life, perhaps five years, but might be renewed if progress is clearly demonstrated. The evaluation function, however, should continue. In addition, a government-wide program of awards for fostering new technology (similar perhaps to the Malcolm Baldridge Award for quality) should be established, perhaps within the Department of Commerce or the Office of Science and Technology Policy, but explicitly coordinated with the evaluation and demonstration activities already recommended.

ACTIONS BY FACILITIES AGENCIES

Agencies responsible for the design and procurement of facilities must adopt a long-term commitment to the enhanced performance and productivity that new technology can yield. These improvements in performance and productivity may be realized at any stage of a facility's service lifetime and consequently may require higher or lower initial expenditures, compared to conventional practice. Care must be taken to ensure that funds needed to achieve longer-term objectives are not diverted when apparent short-term savings result from applications of new technology.

Each such agency should give particular attention or **credit in** the design and construction **procurement** process **to bidders or contractors who propose to apply potentially effective new technology**. Proposed new technologies should have been developed to the stage of prototypical application but need not have been applied in practice.

Agencies should use in construction, maintenance, and repair procurements, **performance specifications which encourage proposers to offer new technology that may not meet more traditional standard specifications.** Although such specifications may sometimes be less explicit than traditionally trained procurement officials might wish, careful technical reviews can be used to evaluate competing offers and ensure that the bases for decisions are adequately documented.

Agencies should **increase integration in the facility design-procurement-construction process**, both to encourage innovators to apply new technology at any stage of the facility life cycle and to permit them to capture the benefits of innovation. Greater use of alternative procurement mechanisms, such as design-build or build-operate-transfer should be encouraged to promote such integration on a broad scale. More comprehensive design contracting (e.g., including interior design with basic architecture) is a more limited tactic that can also be used. Changes in contracting methods to encourage longer-term relationships between contractor and agency, perhaps including multiyear procurements to fund innovation, can also facilitate applications of new ideas.

Each agency should **reward efforts to innovate**. Agencies can establish programs to promote projects showcasing new technology or establish design competitions based on applications of new technology. Agency design awards programs should include innovation as an evaluation criterion. As already proposed, special government-wide awards for fostering new technology should also be established. Awards citations should highlight the contribution of new technologies to agency effectiveness.

ENABLING ENVIRONMENT

Executive and legislative oversight agencies must establish the supportive environment that will enable—and encourage—agency professionals to act aggressively to foster new building technology. Resources and leadership are needed.[36] **Adequate budgets must be made available** to pay the added costs of planning and design analyses that may be required, and responsible officials must have adequate time to oversee projects on which new technology is being applied. In addition, **senior agency and congressional officials must accept that new technologies sometimes may not perform as expected.** Programs to apply new technology need effective management, but must allow for the uncertainties and failures that may accompany such experiments. Systems are needed to **offer insurance and indemnification for both providers and users of new technology.**[37]

Although research and development do not necessarily lead to innovation, the large scale of the building-related industries suggests that small increases in overall productivity will justify limited increases in spending for research, development, and technology transfer. The structure of the industry makes it appropriate that this spending be initiated as a government program activity. In these times of government deficits, substantial spending increases are unlikely even if their payoffs would be substantial, but a relatively modest increase (perhaps 10 to 15 percent) in the estimated $200 million to $230 million spent annually for federally supported building research, development, and demonstration may be possible and could produce significant results,[38]

---

[36]A good example is the Ohio Infrastructure Institute. This coalition of nine colleges of engineering in the state, in partnership with municipal and county engineers, the Ohio Department of Transportation, and design and construction companies, was established to bring about innovation in the design, rehabilitation, and maintenance of public works; to test unproven technologies; and to transfer technology directly to the practitioner. The institute receives some federal government research grant support but could work more actively as a partner with the federal government.

[37]The Civil Engineering Research Foundation, for example, is promoting the formation of joint groups of manufacturers, constructors, insurance companies, and design and legal professionals to review, endorse, and then support applications of promising new technologies.

[38]For example, if the increased spending yields a comparable increase in productivity growth in the federal construction sector, (e.g., an added 0.5 percent increased productivity growth annually), then the return on $20 million to $30 million investment could exceed approximately $75 million for the federal

particularly if the spending is matched by private sector contributions and concentrated in a relatively few programs and major projects. **The building-related industries should petition Congress to establish a program of integrated research, development, and incubation of new building technology to support the establishment or continuation of several "building technology centers of excellence."**

The program, funded at perhaps $20 million to $30 million annually, might be administered by the National Science Foundation as part of the existing engineering research centers program, but might most effectively be assigned to a mission-oriented agency with broad construction responsibilities. Centers—combining participation of industries, universities, state and local governments, and possibly federal laboratories, and based at any of these locations—would be defined to concentrate on specific themes of broad industry significance, such as development of advanced structural concepts, advanced manufacturing and fabrication technologies, or building environmental control. Federal funds would be matched by other participants in the centers' programs.

These centers might also play a role in testing and verification of new technology. Participation of the insurance industry in these centers might facilitate realistic assessment of the risks inherent in any particular technology, as well as more effective sharing of this risk among those who gain from building-related innovation.

TECHNOLOGY TRANSFER

Committee members noted that the transfer of new technology from research to practice is a critically important problem in the building-related industries (and other industries as well). A major objective in virtually all of the committee's recommendations for implementing a more effective government role in fostering new building technology is enhancing technology transfer. The committee agreed that more effort in this area is warranted, particularly with regard to the practices and accomplishments of the federal laboratories involved in building-related research.

The Building Research Board thus plans to conduct a more thorough review of the experience and current practices of the federal laboratories in terms of technology transfer activities. Laboratories working in other technological areas will be considered as well, to search for transferable lessons and opportunities for cross-fertilization of ideas among diverse researchers. Subject to the

---

government alone. Additional benefits would be distributed throughout the private and non-federal government sector.

availability of funds, this review will be conducted during the latter half of 1992, with Federal Construction Council sponsorship.

INNOVATION AND THE FUTURE

Looking to the longer term, the committee found little cause for optimism. New technologies are emerging that offer opportunities for enhanced service, greater efficiency, and protection of natural environmental resources, but much of the new product and process development seems to be occurring outside the United States. The growth of global markets can enhance productivity and the dissemination of improvements to all people but will continue to place political and economic pressure on U.S. industries. Changing international relationships suggest that U.S. resources formerly devoted to military purposes may be increasingly available to seek improvements in our quality of life, but despite the potentially high and widespread payoffs of greater attention to their promotion and output, the building-related industries face substantial competition for support in public policy.

The prospect is daunting. However, committee members observed that U.S. industry and the nation's research establishment continue to produce a stream of new ideas. These new ideas—new technology—are a resource to be tapped. Improving our ability and willingness to put new ideas into practice poses the greatest challenges to the nation's future productivity and continuing high quality of life.

These challenges must be faced. The risks and inevitable discomforts of change must be managed rather than permitted to block progress. The U.S. building industry is being called on to evolve under conditions of uncertainty. Government agencies, as both users of the industry's products and instruments of national policy, have a role to play. These agencies, by trying new ideas and demonstrating that these ideas can indeed yield benefits of improved productivity and performance, can foster the new technology and innovation on which the future of the industry depends. The committee believes that this role is important to the agencies, the industry, and the nation.

APPENDIX A

BIOGRAPHICAL SKETCHES OF COMMITTEE MEMBERS AND STAFF

DAVID R. DIBNER, Vice President and Principal Architect, Sverdrup Corporation, received his B.Arch. from the University of Pennsylvania and has spent his professional career in the practice of architecture in both public and private sectors. He has been a principal in several architecture/engineering and interior design firms. He is a former Assistant Commissioner for Design and Construction, General Services Administration. He was an adjunct assistant professor of architecture at Seton Hall University. He is a Fellow of the American Institute of Architects and has served as a member of the Building Research Board and as chair on several National Research Council advisory committees.

ROBERT J. BOEREMA is head of the St. Augustine office of the Division of Building Construction of the State of Florida's Department of General Services. He received his B.Arch. from the University of Michigan. His professional career includes service with the U.S. Navy Civil Engineer Corps and 28 years of private architectural practice. He served as chairman of the national American Institute of Architects (AIA) Committee on Public Architecture, and President of the Florida Association of AIA, and is a Fellow of the AIA.

LLOYD A. DUSCHA is a consulting engineer, specializing in engineering and dam management engineering. He was formerly with the U.S. Army Corps of Engineers, serving in various civilian capacities in military construction and civil works programs; at retirement he was Deputy Director of the Engineering and Construction Directorate. He has served as a consultant to the Peoples Republic of China and to the World Bank on water resource projects. He received his B.Civ.Eng. from the University of Minnesota. He is a Fellow of the American Society of Civil Engineers and member of the National Society of Professional Engineers, the International Commission on Large Dams, and the National Academy of Engineering.

MARTHA W. GILLILAND is a Vice Dean of the Graduate College and Assistant Vice President for Research at the University of Arizona. She received a B.A. (cum laude) in geology from Catawba College, M.A. in geophysics from Rice University, and Ph.D. in environmental engineering and systems ecology from the University of Florida. Active in community and professional organizations, she was selected Professional Woman of the Year in 1985 by the Nebraska Women's Political Caucus and was a member of the Nebraska Power Review Board. She serves as a member of the Research and Education Advisory Panel for the U.S. General Accounting Office. Until August 1990, she was Director of the Center for Infrastructure Research and Professor of Civil Engineering at the University of Nebraska at Lincoln.

MICHAEL GREEN, Professor at the University of Iowa College of Law and a member of the bar in Pennsylvania and Iowa, teaches and writes on matters of torts, product liability, civil procedure, evidence, and trial advocacy. He received a B.S. degree (summa cum laude) in electrical engineering from Tufts University and a J.D. (magna cum laude) from the Law School, University of Pennsylvania. Active in a range of community and professional organizations, he is coauthor of a recently published book on products liability and safety and the author of numerous articles in the same field.

DON E. KASH is George Hazel Chair in Public Policy at the Institute of Public Policy at George Mason University in Fairfax, Virginia. He has written extensively on matters of public policy in research and technology, and on public-private cooperation, and served as Chief, Conservation Division, U.S. Geological Survey. He received his B.A., M.A., and Ph.D. degrees in political science from the University of Iowa. Dr. Kash has served on a number of national advisory panels, including (as chair) the Oil Spill Countermeasure Technology Working Group for the Office of Technology Assessment and the Engineering Research Centers Panel for the National Science Foundation.

STEPHEN R. LEE, an educator, researcher, and practitioner with broad experience in both systems built housing and advanced technology, is Administrative Director of the Center for Building Performance and Diagnostics at Carnegie Mellon University and past Director of the Pennsylvania Advanced Technology Housing Consortium. He also maintains a private design practice with Tai+Lee, Architects P.C. He received the B.Arch. and M.Arch. from Carnegie Mellon University, and completed one year of self-supported research on industrialized housing in Japan. He was recipient of the 1976 Prize of the Soviet Union, the 1978 Pittsburgh AIA Design Award, and 1986 Progressive Architecture Applied Research Award.

ALVIN P. LEHNERD is Vice President for Research, Design, and Product Development with Steelcase, Inc. He received a degree in electrical engineering from the Ohio State University and a masters degree from George Washington University, and has held manufacturing, product development, and senior management positions with several of the nation's leading corporations. He has lectured widely at universities, corporations, and professional and economic development organizations; sits on the Johns Hopkins Engineering School Advisory Board; and has served on a number of other university advisory boards. Mr. Lehnerd recently served on an advisory panel for an Office of Technology Assessment project entitled "Technology, Innovation, and U.S. Trade."

PHILIP B. LOVELL, an Operations Manager for Turner Construction Company, currently based in their Seattle, Washington office, has had 25 years of experience in all diversity and phases of construction management with Turner. He has completed preconstruction and construction experience in such areas as industrial, medical, biomedical research, cleanroom manufacturing, housing, lifecare, and athletic facilities, as well as other varied institutional and commercial projects. He received a B.S. from Trinity College and a B.S.C.E. from Columbia University School of Engineering and Applied Science, Mr. Lovell is a licensed builder in the city of Boston, and a registered professional engineer in Massachusetts and Connecticut.

FRED MOAVENZADEH, an internationally recognized authority on construction materials, the construction industry, and technology and development, is the George Macomber Professor of Construction Management, and Director of the Center for Construction Research and Education at the Massachusetts Institute of Technology. He received a B.S. degree from Teheran University, an M.S. degree from Cornell University, and a Ph.D. from Purdue University. He is active in a number of professional societies.

TOM F. PETERS, recognized internationally as an educator, lecturer, researcher, author, and historian, is Director of the Institute for the Study of the High-Rise Habitat—Building and Architectural Technology Institute and Professor of Architecture and History at Lehigh University. He received his M.Arch. and Dr.Sc.Techn. from the ETH Zurich. He has served on national and international professional and academic committees, notably, as chairman of the Associated Collegiate Schools of Architecture fifth, sixth, and seventh National Conference on Technology. Dr. Peters has written widely on subjects of the history of building technology, both in technical journals and in more popular magazines such as *American Heritage of Technology and Invention.*

ROBERT T. RATAY is Industry Professor of Civil Engineering at the Polytechnic University in Brooklyn and Farmingdale on Long Island, New York and a consulting engineer in private practice. He was formerly Dean of the School of Engineering at Pratt Institute. The recipient of B.S. and M.S. degrees in civil engineering and a Ph.D. in structural engineering from the

University of Massachusetts, he has practiced for 28 years as a designer and consultant to leading design and construction firms and legal practices on projects throughout North America, Europe, and the Middle East. He has worked, published, lectured, and chaired technical committees on matters of structural safety and forensic engineering. He is a licensed professional engineer in New York and four other states, a Fellow of the American Society of Civil Engineers, and a founding member of the National Society of Architectural Engineers.

JOHN W. THOMPSON, Assistant Vice President, Facilities Planning and Development, for CIGNA Corporation has more than 25 years of experience managing facilities and building programs for large public and private enterprises. He received a B.S. in civil engineering from the University of Connecticut. He is a member of the Executive Board of the University of Hartford Construction Institute and is active in community groups developing low-cost housing in Hartford and other Connecticut communities.

LOUIS G. TORNATZKY is Scientific Fellow, Office of the President, and Adjunct Professor, Department of Psychology, at the Industrial Technology Institute. A noted author, educator, and practitioner in the area of technological innovation, he received his B.A. in psychology from Ohio State University and Ph.D. from Stanford University. He has served on advisory committees addressing such diverse interests as engineering management, knowledge transfer, community health care, and education. He is a fellow of the American Psychological Society and a member of the IEEE Engineering Management Society.

Staff

ANDREW C. LEMER is an engineer-economist and planner. Formerly division vice president with PRC Engineering, Inc., Dr. Lemer is founder and president of the MATRIX Group, Inc., and has written widely on matters of infrastructure, building economics, and development policy, often in conjunction with his work on major projects in the United States and overseas. He received his S.B., S.M., and Ph.D. degrees in civil engineering from the Massachusetts Institute of Technology and is the recipient of a Loeb Fellowship at Harvard University Graduate School of Design. He is a member of the American Institute of Certified Planners, the American Society of Civil Engineers, the Urban Land Institute, and the American Macroengineering Society.

APPENDIX B

PROCESSES OF TECHNOLOGICAL INNOVATION

Louis G. Tornatzky and Andrew C. Lemer

The study of technological innovation is a diverse and growing field. Terminology and theories describing the factors influencing the production and application of new technology differ among observers and researchers in the field, and few studies specific to building technology have been made. The committee undertook a brief review of the field to provide a common basis for its discussions.

In the most general terms, innovation is the introduction of a new idea (Mish, 1985). This introduction entails the production of new information and the diffusion of that information to people who can use it to solve problems, to see the world in a new way, or to enhance their efficiency, effectiveness, or living quality.

In a more specific application, *technological innovation* refers to the process in which a new idea is embodied in tools, devices, or procedures that are of practical value to society. Typically thought of as a new product, technological innovation may also be a new process of production; a substitution of a cheaper material, newly developed for a given task, in an essentially unaltered product; or the reorganization of production, internal functions, or distribution arrangements, leading to increased efficiency, better support for a given product, or lower costs.

Technological innovations often involve both tools and procedures, products and processes, interacting in new ways. Known drugs may be found to be successful in treating new illnesses, or changing the production line may yield improved rates of production. Many of the construction industry's technologies involve such combinations of hardware and software.

Technological innovation can also be an improvement in instruments or methods of making or doing innovation (Kline and Rosenberg, 1986). Industrial research facilities such as the telephone industry's Bell Labs have been recognized as major contributors to innovation in electronics.

New technology that is not used is not innovation. Paradoxically, even technology that is well known and widely used in some industries or nations may still be new and innovative in a different setting. Many years can sometimes be required for new ideas and information to diffuse from one place or application to another. Such technology is still "new" to the society that receives its benefits. Although many people have come to regard new technology solely as the result of increasingly revolutionary discoveries in science and in our understanding of how things work, adaptations and new applications of older knowledge may also lead to innovation.

Successful new technology and innovation tend to be inspired by the practical needs of individual people or enterprises, or the needs of many individuals expressed in market demand or social policy. Technological innovation may also be initiated by scientific invention—new discoveries and developments—but "market pull" is widely felt to be more influential than "technology push" as a force for innovation. The time between invention and innovation may be long.

Whereas technological creativity tends to be "down-to-earth, with such mundane characteristics as dexterity and greed at the center of the act" (Mokyr, 1990), it shares with other forms of creativity an "occasional dependence on inspiration, luck, serendipity, genius, and the unexplained drive of people to go somewhere where none has gone before." The climate within which this creativity can occur, and innovation flourish, is said by many observers to be fragile and highly sensitive to social and economic conditions. On the whole, the forces opposing technological progress have been stronger than those striving for changes, and the study of technological progress is therefore a study of cases in which rare circumstances have permitted "the normal tendency of societies to slide toward stasis and equilibrium" to be broken (Mokyr, 1990).

Innovation is not necessarily good, and in any case, is of little value for its own sake (see box). Rather, innovation is an instrument to achieve broader goals of improved economic productivity, stronger competitive stance in international markets, and improved quality of life. Neither organizational theory nor empirical research supports the notion that innovative individuals or groups will unequivocally be more productive (Tornatzky, et al., 1990). Nevertheless, the absence of innovative adaptation to an environment characterized by rapid change is a reliable indicator of future decline and possible extinction for economic enterprises as well as biological species.

In general, technological innovation takes work (see box). Invention may spring from either focused action or accidental discovery, but innovation requires conscious effort to apply new technology. The motivation for that

INNOVATION IN PERSPECTIVE

In this technological age, innovation is often cited by the popular press and public policymakers as an essential element of our international competitiveness and quality of life. However, by simple definition, innovation is simply the introduction of something new—a new idea, method, or device—and it is not necessarily clear, as the term is commonly used, that innovation is good and desirable. Further, if innovation is not necessarily good, what are the indicators or measures of "goodness" that may be applied to judge whether a new idea, method, or device is a valuable innovation?

The economic literature generally uses growth in productivity as a measure of technology's contribution to an activity. New ideas and methods termed "innovative" tend to be those felt to improve productivity, either by reducing the resources required to accomplish some end or by facilitating the production of new goods or services. Many people would probably acknowledge radar and computer-aided design systems as innovations, while few would readily do so for hazardous wastes associated with construction.

The judgment of value may sometimes seem ambiguous: The year 1915 marks the introduction of both Einstein's general theory of relativity and processed cheese. Bulliet attributes to the invention of the camel saddle, sometime between 500 and 100 B.C., the camel's gradual displacement of wheeled transport in the Middle East and a subsequent lasting bias that was still visible centuries later in the dearth of wheelbarrows on construction sites in Tehran (Bulliet, 1975).

conscious effort is typically economic. Historically, Western technology has developed primarily in an economic context and has often been regarded as merely an outgrowth of economic needs and institutions (Rosenberg and Birdzell, 1990).

Economic and technological factors are intertwined, perhaps inextricably, in the innovation process, but the possibility of achieving improved safety or other benefits not immediately measured in monetary terms often provides the incentive for innovation. The committee notes that lessons from studies of innovation in several fields suggest that users' needs are an important source of innovation.

WORKING FOR INNOVATION

In the beginning there is just you, a field, and some lettuce seeds. You prepare the field, sow the seeds, tend the plants and reap 20 heads of lettuce. Then from the sky falls a rake. There is still you, a field and some lettuce seeds, but using the rake has allowed you to double your production.

That is the free lunch provided by technology.

So said Susan Lee (1991) in her review of Joel Mokyr's book on technological creativity and economic progress, *The Lever of Riches*.

But Mokyr argues there is no free lunch. According to Mokyr (1990), technological creativity shares with other forms of creativity an "occasional dependence on inspiration, luck, serendipity, genius, and the unexplained drive of people to go somewhere where none has gone before." People striking out into uncharted territory discover new ideas, and innovation results when people work and take the risks to put these new ideas into practice. "Sustained innovation requires a set of individuals willing to absorb large risks, sometimes to wait many years for the payoff (if any). It often demands an enormous mental and physical effort on the part of the pioneers."

Mokyr suggests that risk aversion, leisure preference, and time preference influence the willingness of people at any particular time and place to make this effort, and these factors are thus of major importance in determining the rate of innovation in a particular society.

DESCRIBING INNOVATION AS A PROCESS

Technological innovation has been described often as a linear process of distinct stages or phases: innovation begins with scientific discovery, proceeds through development of practical applications of this discovery, and finally achieves success as dissemination and implementation at the hands of users (see **Figure B-1**). This linear model is overly simplified. In fact, the innovation process may be quite nonlinear, drawing repeatedly on basic knowledge, responding to newly perceived needs, and modifying earlier concepts of the tool, device, or procedure that eventually evolves (Tornatzky et al. 1990). Nevertheless, the progress of innovation requires, first, understanding of the basic principles and processes that permit manipulation of the physical environment, and then the interaction of often complex social forces through which this understanding is to be put to use.

Innovation overall may occur through the effect of small advances that cumulatively prove decisive in productivity growth or through great leaps of discovery that represent radically new ideas without clear precedent. Some argue that almost all innovation is a result essentially of the former, in that every major invention is followed by a period of learning and application that accounts for the bulk of growth. In any case, the large discoveries and small steps of exploration are complements, not substitutes (Mokyr, 1990). Solving seemingly mundane problems requires real creativity and can produce big payoffs.

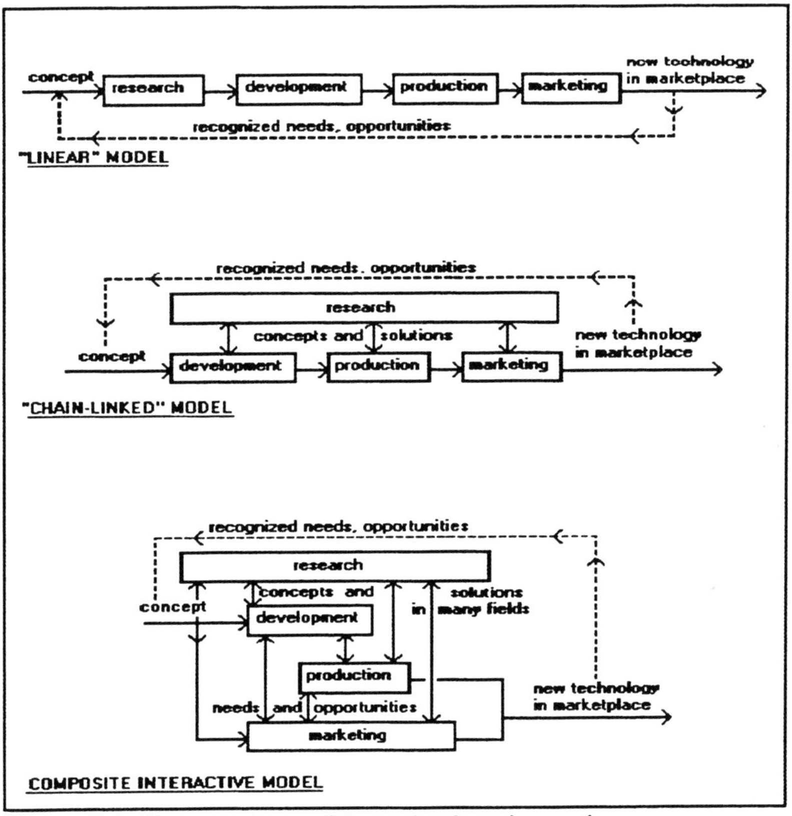

Figure B-1 Alternate views of the technology innovation process.

NEED IS THE STARTING POINT

In almost all cases of successful innovation, some unmet societal need is in the mind's eye of the innovator. However, in the quest for understanding of technological innovation, it is inventions, whether in small steps or great leaps, that have been the focus of illuminating case studies. Few generally accepted lessons about how invention occurs have emerged, but most of these stories share one important characteristic: a person or team of people, intimately familiar with both the new technology at hand and the potential for gain in applying that new technology, is pivotal (Kash, 1989).

Sometimes the invention or new idea occurs to an individual or small group, such as Henry Ford's creation of a mass production assembly line for automobile manufacturing or the invention of nylon in a corporate research laboratory. At other times, the new idea emerges in a more diffused fashion, and it may even be difficult to identify precisely what the invention is. The constant search for improvement (termed *keizen* in Japanese) is such a diffuse source, producing apparently substantial benefits for Japan's auto industry. Generally speaking, invention is more likely to be encouraged when the inventor can capture the benefits of his or her work through licensing fees, product sales, or fees for services.

Integrated consideration of design and production concerns seems to favor innovation. Design has been defined as the "process of applying scientific and technical principles to meet requirements for suitable arrangement, appropriate use, convenience, ease and economy of manufacture, and acceptable appearance" (Pye, 1964).

As practiced in industry, particularly with regard to consumer products, design is closely tied to the production and marketing efforts that lead to commercial success and the broader adoption of new ideas in the marketplace. Such close ties are the exception in most segments of the building industry, and the typical separation between designer and construction contractor hinders innovation.

DIFFUSION OF NEW TECHNOLOGY INTO USE

After invention has occurred, the new technology must enter practice to become effective innovation. If successful, the new idea may spread or be communicated to other users. New ideas may spread to other fields as well, spawning subsequent generations of innovation. For example, flat cable technologies developed initially for aerospace applications later became innovations in building controls and office automation.

An important aspect of this spreading is incorporating the user's perspective in the new technology's application. Few technologies are "self-executing" in the sense that users can readily understand how to adopt them effectively to achieve benefits. Technologies that are particularly complex, different from those currently used, or costly to adopt may call for considerable adaptation and accommodation by the users. This is especially so when the user is a group or organization, rather than an individual (see **Figure B-2**); (Tornatzky et al., 1982; Tornatzky et al., 1990).

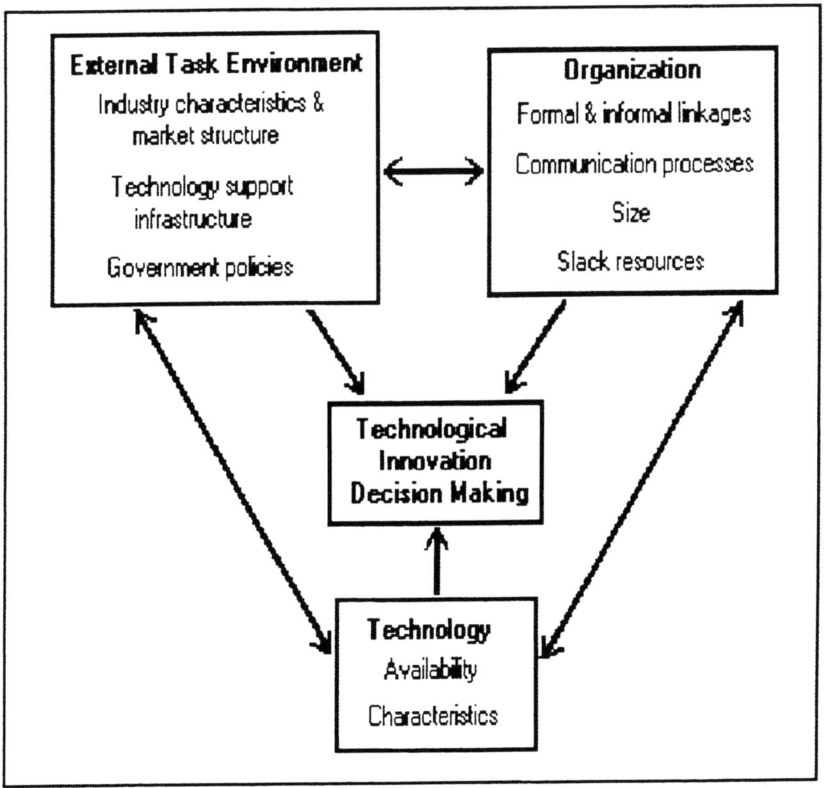

Figure B-2 The context for adoption of new technology (from Tornatzky et al, 1990).

The spreading of innovation into broad practice is frequently termed *discipline* and has become a subject of much study (Rogers, 1986). The study of diffusion as a discipline evolved after World War II, spreading through application in such fields as agriculture, medicine, education, transportation, and others.

The study of diffusion suggests (Grubler, 1990) that innovations, even those embodied primarily in hardware and production techniques, are intrinsically interrelated to organizational and social adaptation processes. Once innovations appear and demonstrate potential technical and economic viability, they are put forward for societal "testing." They are either refused or accepted and, if accepted, begin to spread and interact with existing techniques that satisfy the same human need. Innovations that prove to be better adapted to the technical, economic, and social requirements imposed by society and its economy will gradually replace existing techniques and practices. When a distinct effort is made to encourage diffusion of new technology, particularly new technology resulting from the discoveries and inventions of a particular institution, the effort is often termed *technology transfer*.

Gatignon and Robertson (1986) propose, from a marketing perspective, five factors influencing the filtering and persuasive effect of information transfer: (1) availability of positive information (negative information has much greater impact); (2) credibility of information (viewed as "objective" or from influential sources); (3) consistency of information (greater consistency has higher impact); (4) type of information source (media or personal contact; the latter is more influential); and (5) personal characteristics of the individuals involved in the process. For example, initial negative experience with digital Heating, Ventilating and Air Conditioning controls in U.S. Army installations, attributable in substantial measure to inadequate training of maintenance personnel, has made it very difficult to consider such devices in current military construction. Formal programs of technology transfer that depend primarily on written materials are often less productive than those that encourage direct and frequent contact between the sources of new ideas and the potential users of those ideas.

ROLE OF RESEARCH

Research, a conscious and directed effort to develop new things, may not be necessary to innovation, but it facilitates the process. The early and overly simplified "linear model" of innovation described the bringing of new ideas into use as a progression from research to development to production to marketing. Although the linear model is still used in discussion, many investigators agree that it should be consigned "to the scrap heap of history" (Ziman, 1991), and alternatives have been proposed to reflect better the complex interactions of researchers and users of research (see **Figure B-1**). The "chain-linked model" focuses more on the potential market demand for an innovation as a motivator of invention or design. Research in this latter model is an ongoing stream of activity in parallel with product development, production, and marketing.

Research is seen to contribute, in principle, to all other stages of the process (Kline and Rosenberg, 1986).

Other models incorporating feedback from later stages in the sequence to earlier ones are said to reflect the real relationships that operate in a major corporation and within the community of scientists and engineers seeking new technology. Some analysts have also tried to consider the ways in which research and development activities in other fields can spur innovation, as reflected in the "composite model" illustrated in Figure B-1. One writer suggests that the sources of innovation are better comprehended as nodes in a multilayered and interconnected "neural network" that includes many diverse ideas and disciplines (Ziman, 1991).

Some observers assert that the contributions of science to economic growth and industrial technology began in the late eighteenth century. The pressures for economic gain—through exploration and industrialization—drove the engineering innovation of that period that underlay much of Europe's Industrial Revolution. The success of efforts to explain natural phenomena with theory inaccessible to those who lack special training, and the creation of industrial research laboratories capable of extending theory, have brought science into the economic sphere and made its advance inseparable from that of industrial technology in Western economies (Rosenberg and Birdzell, 1990).

The value of research as a source of innovation is difficult to estimate, as is the likelihood that innovation will occur under any given set of circumstances. It is reported (Rosenberg, 1986), for example, that Charles H. Duell, then commissioner of the Patent and Trademark Office, recommended to President McKinley at the beginning of the twentieth century that the agency should be closed down, because "everything that can be invented has been invented." This extreme example illustrates the persistent underestimation of future technological change.

However, experience shows (e.g., Mansfield, 1968) that technological innovation draws on the fundamental knowledge produced by research. Moreover, as the technological content of new products and processes increases, the relationship between innovating organizations and basic science research becomes more active. Such observations are strong circumstantial evidence that research is a solid contributor to technological innovation, and some writers suggest that the industrial research laboratory, specifically established to facilitate exploitation of scientific knowledge for industrial purposes, is "one of the most important institutional innovations of the twentieth century" (Rosenberg, 1986).

The contribution of research to innovation may be limited by the characteristics of the people involved in both the research and its application. Technical specialists seem to be typically capable of extending and improving methods of their own expertise and applying them to new uses. Any competent specialist is then likely to be reasonably good at anticipating the kinds of performance improvements that can be teased out of a given technology. This capability may well be the primary basis for innovation in the construction industries, which traditionally occurs primarily on the job site.

However, the very nature of an expert's education and professional experience is likely to disqualify that person from developing very new technologies based on different principles or even from appreciating the potential significance of new principles. Forming cross-functional or multidisciplinary teams that bring together individuals with differing perspectives is one means for overcoming this limitation. The individuals in such teams may see new ways of applying the principles and practices that, to their associates in other fields, are standard and lacking in the potential for innovation (National Research Council, 1991).

REFERENCES

Bulliet, Richard W. 1975. The Camel and the Wheel. Cambridge Mass: Harvard University Press.

Gatignon, H. A. and T. S. Robertson. 1986. Integration of consumer diffusion theory and diffusion models: New research directions. In Innovation Diffusion Models of New Product Acceptance, V. Mahajan and Y. Wind, eds. Cambridge, Mass.: Ballinger.

Grubler, A. 1990. The Rise and Fall of Infrastructures: Dynamics of Evolution and Technological Change in Transport. Germany: Physica-Verlag Heidelberg.

Kash, D. 1989. Perpetual Innovation. New York: Basic Books.

Kline, K., and N. Rosenberg. 1986. An overview of innovation. In The Positive Sum Strategy: Harnessing Technology for Economic Growth, National Research Council, 1991, Landau, R., and N. Rosenberg, eds. Washington, D.C.: National Academy Press.

Lee, S. 1991. Technology: A free lunch for all. Wall Street Journal, March 4, A7.

Mansfield, R. 1968. Industrial Research and Technological Innovation: An Economic Analysis. New York: W.W. Norton.

Mish, F. C., ed. 1985. Webster's Ninth New Collegiate Dictionary. Springfield, Mass.: Merriam Webster.

Mokyr, J. 1990. The Lever of Riches: Technological Creativity and Economic Progress. New York: Oxford University Press.

Pye, D. 1964. The Nature of Design. London: Studio Vista Press.

Rogers, E. M. 1986. Three decades of research on the diffusion of innovations: Progress, problems, prospects. Paper presented at the Conference on Innovation Diffusion, Venice, March 18-22.

Rosenberg, N. 1986. The impact of technological innovation: A historical view. In the Positive Sum Strategy: Harnessing Technology for Economic Growth, Landau, R., and N. Rosenberg, eds. Washington, D.C.: National Academy Press.

Rosenberg, N., and L. E. Birdzell, Jr. 1990. Science, technology and the Western miracle. Scientific American 263(5):42-54.

Tornatzky, L. G., and M. Fleischer, et al. 1990. The Process of Technological Innovation. Lexington, Mass.: Lexington Books.

Tornatzky, L. G. and K. Klein. 1982. Innovation characteristics and innovation adaption-representation: A meta-analysis of findings. IEEE Transactions on Engineering Management (EM29):28-45.

Ziman, J. 1991. A neural net model of innovation. Science and Public Policy 18(1):65-75.

APPENDIX C

REVIEW OF SPECIFIC AGENCIES' STANCE TOWARD BUILDING INNOVATION

Federal agencies are not strictly subject to state and local government building regulations, although federal law encourages them to conform to applicable regulations in jurisdictions where federal projects are located. The agencies have adopted their own design guidelines and specific criteria that in many cases are identical to those contained in model codes and in the state and local regulations based on these codes. Nevertheless, federal agencies have an authority in principle to adopt design practices, construction procedures, or new technologies that are not accepted under state or local regulations. In practice, this authority may be constrained by public opinion and agencies' aversion to the potential losses when new technology does not perform as hoped.

Liaison representatives of agencies sponsoring the study presented to the committee the current concerns and interests of their agencies in matters of new technology in the building industries. Although the sample of agencies surveyed is necessarily limited, the concerns and interests raised are a representative cross section of issues regarding government's role in building innovation.

NAVAL FACILITIES ENGINEERING COMMAND

The leadership of the Naval Facilities Engineering Command (NAVFAC), among the principal motivators of the current study, finds itself called upon to demonstrate new building technology that goes beyond the agency's basic mission. A case in point is the application of seismic base isolation devices developed by a private entrepreneur but not yet demonstrated in U.S. applications. The agency lacks adequate funds for product testing and often finds congressional oversight unforgiving when new technology fails to perform as promised. Furthermore, personnel as well as funding tend to be fully occupied with the day-to-day demands of the agency's mission, which does not seem to include explicit responsibilities for furthering innovation. NAVFAC leadership thus questions whether it is appropriate for the agency to assume the risks of trying new technology.

Although the agency may benefit from a particular application—and, when the benefits seem likely to outweigh the risks, will opt for the new technology—it is the private developer of that technology who reaps the larger commercial benefits of successful application. Liability issues are not generally a constraint on decision, although the agency must typically relieve the architect/engineer and constructor of liability when new technology is applied. Rather, the question for NAVFAC is whether broad national interests in innovation should influence individual project decisions in ways that may in some instances pose risks to the agency's effective and efficient performance of its basic mission.

U.S. ARMY CORPS OF ENGINEERS

The U.S. Army Corps of Engineers' current spending for buildings and other facilities includes new construction, rehabilitation and retrofit, and maintenance for other agencies as well as the Army. In its own construction the Corps faces many of the concerns expressed by NAVFAC. The Corps has notable examples (e.g., early experience with heat pumps for air conditioning) in which overly aggressive efforts to adopt new technology led to failures that actually delayed innovation.

However, the Army has been assigned a more active role in fostering the development of new technology. The Construction Productivity Advancement Research (CPAR) Program, initiated in 1989, is a cost-shared research, development, and demonstration program, under which private enterprise and the Army work jointly to develop new ideas within a limited range of technical areas consistent with the Corps' primary mission. Examples of new technology developed under the program include a mechanical device to assist masons in lifting heavy blocks, and computer programs to assist facilities designers and managers to perform life-cycle economic analyses of major building subsystems.

The Army has several research facilities that support its missions and provide technology support to the construction sector. The CPAR program and other broad federal legislation have placed considerable emphasis on technology transfer, and the Army has capitalized on it via an aggressive Facilities Engineering Application Program (FEAP) and the Cooperative Research and Development Agreement (CRDA) program. FEAP has facilitated the adoption of 120 technologies in the day-to-day operations, design, maintenance, and construction of Army facilities and of 40 technologies into the private sector for marketing to the Army and other customers. The Army has capitalized on the royalty provision of the federal legislation, with $150,000 having been realized. The Army has a continuing research program in technology transfer for the construction industry supporting the Army.

DEPARTMENT OF ENERGY

The Department of Energy (DOE) builds facilities for its own use, and also has active programs of research and technology transfer to foster energy conservation and shifts toward greater use of renewable resources throughout the building sector. The experience with these latter programs provides a number of examples of specific new technologies that have had varying levels of success in being adopted in practice, some of which have been described in Chapter 2.

The National Competitiveness Technology Transfer Act of 1989 has been a factor in shaping DOE programs, and transfer of government-funded technologies to the private sector is considered a part of the department's mission, sought as a means of enhancing U.S. competitiveness. In dealing with this aspect of its mission, DOE must make decisions regarding allocations of effort among the stages of innovation from the discovery of new ideas to putting those ideas into practice.

DEPARTMENT OF STATE

The State Department's embassy construction program presents a wide range of very challenging technical problems, but the department also needs safe, functional, and attractive facilities to house a range of more mundane activities at reasonable cost. The objectives set for facilities of both types are often in conflict (e.g., fire safety, which requires easy access and egress, versus security, which requires that access be strictly limited) and the department finds that decisions about hardware technology may often be made without adequate consideration of the "software" or management issues involved.

For example, a pre-engineered concrete office building was designed for use in New Guinea. The factors motivating this new design included ability to manufacture the parts with better control than could be achieved locally, as well as shortages of local materials. However, the manufactured parts could not be effectively transported to sites in New Guinea, and the new technology was judged inappropriate.

The State Department feels a need for better procedures and criteria for assessing new technology ideas, in view of the conflicting objectives of various participants in the decision-making process. Although the department is fairly effective in learning from previous experience, the management system deals poorly with changing priorities, such as the shift since the 1960s, when architectural design excellence was a primary objective, to a current emphasis on security as the most important factor in design. Such shifts are to be expected during the anticipated long service life of new technology. Such change can pose risks that commitments (e.g., to maintenance or to priorities for a particular energy source) implied in the initial decision will be superseded.

U.S. POSTAL SERVICE

The U.S. Postal Service is relatively unique as a government agency in that its programs are judged somewhat like a private enterprise. New technology can be evaluated in terms of its likely ability to reduce costs or increase revenues.

The Postal Service constructs facilities in three primary size categories, from small post offices in typically rural settings to large and highly mechanized processing facilities. Rehabilitation of older buildings is seldom as cost effective as the construction of new ones that take full advantage of the most current technology (particularly in mail handling). The Postal Service maintains an active search for useful new technology, but the technology must be clearly cost-effective before it is adopted. Professional staff undertake to scan selected literature for new ideas and maintain an informal index card system for capturing these ideas.

Recent examples of successful introduction of new technologies include use of computer simulation for the development of more effective HVAC design principles for 24-hour facilities, applications of postoccupancy evaluation, introduction of lightweight plastic doors, and work on the "store of the future" (prototype retail postal unit). However, identifying cost-effective new technology to fill definite needs is still a major challenge. Concerns for worker safety, and to a lesser degree security of mail, necessitate thorough evaluation of new technology and represent the most significant constraints to the adoption of new ideas.

NATIONAL INSTITUTE OF STANDARDS AND TECHNOLOGY

The National Institute of Standards and Technology (NIST) is a part of the Department of Commerce (DOC), with wide-ranging programs in many areas of science and technology. NIST's Building and Fire Research Laboratory

(BFRL) conducts a broad program of research activities, funded approximately 52 percent by requests from other agencies and 48 percent by directly appropriated funds. The laboratory's aims are primarily to develop technologies to predict, measure, and test the performance of construction and fire prevention and control products and processes, and to foster the exchange of knowledge. The institute undertakes no direct construction.

The system for selection of research topics is based largely on NIST review, congressional mandates, DOC programs, and recommendations of committees of experts. Like the USA Construction Engineering Research Laboratory, the BFRL finds that communication of new ideas resulting from the agency's work to those who might benefit from the knowledge, is a continuing challenge.

The Trade Act of 1988 established a position of Under Secretary for Technology in DOC and a program at NIST to encourage commercialization of new technology developed in the government's labs. NIST's Advanced Technology Program provides partial funding and technical assistance to encourage the development of precompetitive new technology to reduce some of the risks inherent in the application of such technology. Proposals for the support of new building technologies are eligible for assistance under this program.

STATE OF MARYLAND

Like many other state and local jurisdictions, Maryland's primary interests in new technology emphasize maintenance and repair of existing facilities. Past efforts to try new technologies have had a generally low success rate, and efforts to use single-ply roofing are an outstanding example for state officials of the risks of trying new technology. Problems have been encountered in both installation and maintenance.

The state experiences persistent difficulty in recruiting, training, and retaining qualified people for operation and maintenance of new technologies. These difficulties—and the overall challenges of follow-through on action needed to ensure success in the adoption of new building technology—make it difficult for professional staff, who might otherwise be inclined to try new technology, to justify such decisions to elected officials.

MONTGOMERY COUNTY, MARYLAND

Montgomery County, a part of the Washington, D.C., metropolitan area, has over the past decade experienced rapid growth necessitating substantial construction. The county has tried new technologies, particularly for energy management, and found that its biggest challenge is in getting the new technology incorporated early enough in the design process to ensure effective implementation. Securing sufficient funding to permit adequate analysis of new technology options in the planning and design stages of project development has been difficult.

One area in which the challenge has been met, with substantial success, is energy efficiency. County legislation has made energy-efficient design a mandatory part of all county construction. Each request for design services specifies analysis models to be used in the evaluation of energy-saving technology options. The county finds that savings of 30 to 40 percent of future energy costs are achieved with minor increases in design expense—typically less than 0.5 percent of initially estimated total project cost—and no increase in construction cost.

APPENDIX D

FEDERAL LAWS AND REGULATIONS RELATED TO TECHNOLOGICAL INNOVATION IN BUILDING

A variety of federal legislation and regulatory actions have been promulgated over the past several decades, to foster general research and development, advances in specific technology, new special-purpose institutions, and technology transfer from government programs to the private sector. Many of these activities have had relatively direct relevance to the construction industries and facilities development.

National Aeronautic and Space Act, 1958 (PL 85-568). The National Aeronautics and Space Act identified the potential for commercial applications of military and space research as a key mission element of the National Aeronautics and Space Administration (NASA).

Federal Laboratory Consortium for Technology Transfer. Established in 1974, the Federal Laboratory Consortium for Technology Transfer was incorporated in the Federal Technology Transfer Act of 1986 (amended 1989) to promote and strengthen technology transfer across the federal research system. More than 300 laboratories, representing 10 different agencies, comprise the consortium, which supports the technology transfer needs of member laboratories and agencies and of the public and private sectors.

Stevenson-Wydler Technology Innovation Act, 1980 (PL 96-480). The Stevenson-Wydler Act attempted to establish some uniformity of effort among federal agencies, by designating that 0.5 percent of each agency's budget go to an Office of Research and Technology Application (ORTA).

The ORTA was to target technologies that might have applicability elsewhere. The act stated that technology and industrial innovation are central to the economic, environmental, and social well-being of U.S. citizens because they offer (1) an improved standard of living; (2) increased public and private-sector productivity; (3) new industries and employment opportunities; (4) improved public service; and (5) enhanced U.S. competitiveness in world markets.

Patent and Trademark Amendments, 1980. The Patent and Trademark Amendments encourage the licensing of government-operated laboratory inventions by authorizing federal agencies to grant exclusive and partially exclusive licenses. They authorize federal agencies to withhold from disclosure classified information, company trade secrets, and related inventions that are likely to result in a patent application.

Bayh-Dole Act, 1980 (amended in 1984). The Bayh-Dole Act permits independent laboratories run by small or nonprofit businesses doing federal research to retain their right to inventions. (However, agencies have tended to give bids to those laboratories that agree to give up ownership of resulting technology.)

Trademark Clarification Act, 1984 (PL 98-620). Concerned primarily with semiconductor chip products, the Trademark Clarification Act amended the Patent and Trademark Amendments of 1980 by extending coverage to most of the Department of Energy, except laboratories engaged in naval nuclear propulsion or weapons-related programs.

National Cooperative Research Act, 1984 (PL 98-462). The National Center for Manufacturing Sciences (NCMS) was created in 1984. NCMS sponsors and manages the research of member companies and U.S. industries in the area of manufacturing science and technology.

Federal Technology Transfer Act, 1986 (PL 99-502). The Stevenson-Wydler Technology Innovation Act was amended in 1986 to promote technology transfer (TT) by allowing government-operated laboratories to enter into cooperative research agreements, and to designate formally the Federal Laboratory Consortium (FLC) for TT within the National Institute of Standards and Technology. The Federal Technology Transfer Act (FTTA) establishes a TT mission for federal agencies; improves the use of federally funded research and technology by state and local governments, and the private sector; provides federal employees recognition for outstanding contributions to TT; ensures full use of the products of federal investment in R&D; mandates the establishment of ORTAS; and sets aside for TT not less than 0.5 percent of each agency's R&D budget. As implemented by Executive Order 12591, FTTA directs agencies to enter into cooperative research and development agreements with state and local governments, universities, and private companies; awards exclusive licenses for patents to cooperators with federal laboratories; grants awards to federal employees significantly contributing to TT; and implements royalty sharing. The act focuses on small- and medium-sized business, creates a research and technology applications office, and encourages the Department of Commerce to offer expertise on commercial potential of innovations and methods for their commercialism; it provides a minimum 15 percent royalty to federal/nonfederal inventors; and allows the inventor to have exclusive right to the invention. The act has worked well because there is no centralized coordinator for all the agencies.

Facilitating Access to Science and Technology, Executive Order 12591. Executive Order 12591 implements the FTTA by directing the heads of federal agencies to delegate authority to their government-operated federal laboratories to enter into cooperative R&D agreements and to license, assign, or waive rights to intellectual property, which would include inventions and computer software that the laboratories develop. It promotes cooperation among federal, state, and local governments, industry and academia. Federal agencies are directed to improve their TT; the Departments of Agriculture, Commerce, Energy, and Health and Human Services, as well as the National Aeronautics and Space Administration (NASA), are to participate in the technology share program. The President's Commission on Executive Exchange is to assist federal agencies to develop exchange programs for scientists to work in their respective laboratories; and to seek U.S. government technologists to be assigned to U.S. embassies. Similar language is contained in the Trademark Clarification Act of 1984.

Omnibus Trade and Competitiveness Act, 1988 (PL 100-418). The National Institute of Standards and Technology was established by the Omnibus Trade and Competitiveness Act of 1988, along with regional centers for the transfer of manufacturing technology to small- and medium-size firms. Sections 6101 et seq. of the bill established an Officer of Training Technology Transfer in the Department of Energy. The National Technical Information Service (NTIS), under contract, will compile, update, and distribute a government-wide inventory of training technologies.

Training Technology Transfer Act 1988 (PL 100-418). The Training Technology Transfer Act augments federal programs for training new industrial workers and retraining workers displaced by new technologies, and facilitates the transfer of education and training software from federal agencies to the public and private sectors and to state and local governments and universities to support the education and retraining of industrial workers, especially workers in small businesses. The focal point of this activity was placed in the Department of Education's Office of Technology Transfer.

National Technical Information Act, 1988 (PL 100-519). The National Technical Information Act of 1980 amends the Stevenson-Wydler Technology Innovation Act to establish a Technology Administration in the Department of Commerce. It provides for the National Technical Information Corporation to supersede the NTIS. The corporation sets its fees for services. The Secretary of Commerce has responsibility for coordinating the program throughout the Executive Branch. Guest workers at federal laboratories are eligible for royalty sharing.

National Competitiveness Technology Transfer Act, 1989 (PL 101-189). The purpose of the National Competitiveness Technology Transfer Act is to enhance U.S. national security by promoting technology transfer between government-owned, contractor-operated (GOCO) laboratories and the private sector, and to enhance collaboration among universities, the private sector, and GOCO laboratories. It amends the Stevenson-Wydler Technology Innovation Act of 1980. Procedures must be developed that disseminate information on opportunities to participate with the laboratories in technology transfer.

Federal High Performance Computing Program, 1989. The Federal High-Performance Computing Program is coordinated by the Office of Science and Technology Policy to enhance high-performance computing capability. The program is important to the high-technology, small-business segment to give it the tools for accessing cutting-edge technology.

APPENDIX E

NEW BUILDING TECHNOLOGY AND INNOVATION: A SELECTIVE REVIEW

Few studies of the building-related industries include any consideration of changes in technology that have actually occurred over the past several decades. The committee undertook a brief and selective review of these changes to provide a basis for its assessment of the rate of innovation in these industries.

The U.S. Office of Technology Assessment (1987) characterized technological innovation in the construction and materials industries in four categories: (1) development of new technologies within individual firms; (2) application or modification of new technology developed outside of the firm; (3) combining existing technologies in novel ways; and (4) incremental advances in existing techniques. It is generally difficult to distinguish innovation in the second two categories, although members of the committee suggested that these account for the preponderant share of all innovation in building, for both the construction process and the products thereof.

COMPUTER-AIDED DESIGN AND CONSTRUCTION

Much is made of advancements in computer-aided design and computer-aided manufacturing (CAD/CAM) in the automotive and electronics industries. Similar advancements are changing the ways that building designers and constructors do their jobs, and are increasing productivity in the process.

Early progress in the field was spurred by the development, in the 1960s, of the COGO system (named for its ability to handle *co*ordinate *geo*metry) that helped highway engineers locate new routes and structural engineers analyze building frames. The innovation of an effective and—for the time—user-friendly program that solved problems common to almost all parts of the diverse engineering profession spurred development of tools for soil and rock mechanics, steel and concrete structures, highway construction planning, and an array of other applications.

These tools required large computers and programming sophistication that limited their value to those professionals and firms who were sufficiently large or specialized to justify the investment needed to gain access to a system. The general-purpose CAD systems that began to appear in the 1970s, offering more effective data management and faster updating of the huge number of drawings required for the construction of a large facility, had the same problem.

The appearance of powerful desktop computers—the PC (personal computer) and distributed workstations—signaled a change. Programmers made great strides in developing easier-to-use software, and the total investment required to begin using CAD came down to levels that virtually any engineer, architect, or planner could afford.

However, the machines still were tied to the office desktop. Applications in construction were limited primarily to project scheduling and accounting, and to designers' revisions of preconstruction drawings into "as-builts." Now, in the 1990s, the situation is changing quickly.

Constructors are finding that CAD systems allow them to ask "what if" with respect to how hoists, cranes, and other large equipment will fit and infract on the job site. The systems help coordinate trades. They facilitate calculation and control of materials quantities. Powerful portable computers—the new generation of "laptops"—are durable enough for use in the field. With data connections to the central office, site superintendents and field engineers can have full access to all available information about a project.

The new technology, while boosting productivity, is creating what ENR,[39] chronicle of the construction industry's day-to-day activities, termed "cultural chaos." Traditional ways of conducting business among owners, designers, and constructors are changing as constructors acquire design capability, and designers, having already input data for their own purposes, find it easy to make quantity estimates and do other tasks normally left to the constructor. Both groups, as well as owners who could find such data useful in subsequent management of their facilities, are unused to sharing and cooperation. Issues of potential liability and copyright ownership are arising. Those involved in

---

[39]Formerly *Engineering News Record.*

design-construction development, because they create the design files and then use them in-house for construction, avoid some pitfalls and are leading in this broadening application of computers in aid to design and construction.

ENERGY MANAGEMENT IN BUILDINGS

Space heating, cooling, lighting, and other activities make the building sector one of the primary consumers of energy in the U.S. economy. The oil crises of the 1970s and continuing increases in energy process, concerns about global warming and other large-scale environmental effects of energy use, and building owners' ongoing efforts to control operating costs have motivated substantial effort to develop new technologies for the control, conservation, and reduction of energy use.

One study of both gas and electric household appliances shows that the 1985 average efficiency of new appliances purchased, as well as the efficiency of the most efficient new appliances available, are consistently better than the estimated average efficiencies of appliances in service (see **Figure E-1**); (Geller et al., 1987). In many cases, the best available unit is 30 to 40 percent more efficient than the average unit purchased, and current research and development activities promise to reduce unit energy use as much as 40 to 50 percent more during the 1990s.

Space heating and ventilation are major energy consumers in buildings, as well as important factors in determining how well a building is judged to perform. Innovations in these areas thus have both monetary and non-monetary payoffs. The committee conducted an informal survey of federal government mechanical engineers and produced a list of 16 specific innovations that have entered practice in the past 25 years (see **Table E-1**). The committee made a similar survey of government electrical engineers (see **Table E-2**). Two items on the lists are identical (item 3: variable frequency drives, and item 6: energy-monitoring and control systems).

STRUCTURES AND THEIR CONSTRUCTION

While the committee decided to exclude the housing sector from much of its discussion, members noted that the refinement of dimensional lumber in the late nineteenth century, by Bemis and his successors, was a major innovation in home building. Arguments have been advanced to suggest that uniquely American 2 x 4 stud and balloon- or platform-frame building systems are highly flex-

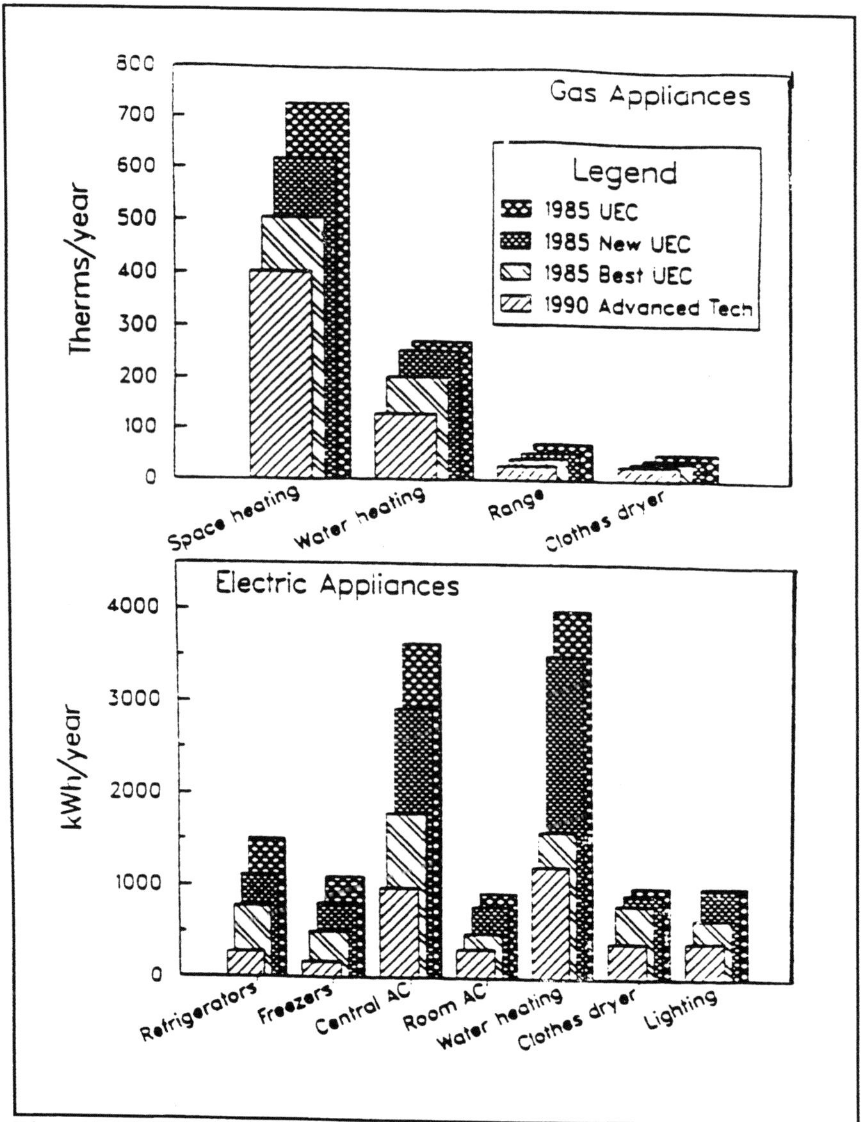

Figure E-1 Unit energy consumption (UEC) of household appliances in the United States, source: Geller, et al., (1987).

Table E-1 Major HVAC Innovations Accepted in the Past 25 Years.

Development	Year Accepted (average of estimates)	Year Accepted (Range of estimates)
1. Screw compressors (replacing some centrifugal and reciprocating compressors)	1969-1970	1965-1980
2. Variable air volume systems (replaced constant-volume systems)	1972-1973	1970-1975
3. Energy monitoring and control systems (central) (new development)	1973-1974	1970-1977
4. Electronic controls (replaced some pneumatic controls)	1974-1975	1965-1980
5. Computer-aided design (loads) (replaced hand calculations)	1969-1970	1965-1972
6. Energy use simulation (replaced degree-day calculations)	1974-1975	1968-1979
7. Variable-frequency speed controls (replaced some other methods of controlling fluid flows)	1983-1984	1978-1987
8. Thermal storage systems (new development)	1980-1981	1966-1988
9. Solar energy systems (new development)	1976-1977	1971-1982
10. Total/selective energy systems (new development)	1980-1981	1976-1986
11. Heat pumps (residential) (new development)	1970-1971	1955-1981
12. Two-stage absorption air-conditioning units (replaced single-stage units)	1977	1965-1982
13. High-efficiency furnaces (residential; replaced low efficiency units)	1983-1984	1980-1986
14. High-efficiency air conditioning units (residential; replaced low-efficiency units)	1982-1983	1980-1986
15. Scroll compressors (replaced some reciprocating compressors)	1985-1986	1984-1988
16. Alternate chlorofluorocarbon refrigerants (replacing some other refrigerants)	1990	1990

Note: Identified at the March 27, 1991 meeting of the Federal Construction Council Consulting Committee on Mechanical Engineering; the committee used a simplified "Delphi" procedure.

Table E-2 Facilities-Related Electrical Engineering Developments Accepted in the Past 25 Years.

Development	Year Accepted (average of estimates)	Year Accepted (range of estimates)
1. Solid-state circuit breakers (replacing other circuit breakers and fuses)	1981	1975-1986
2. Static uninterrupted power supplies (replacing engine-generator sets)	1979-1980	1970-1988
3. Variable-frequency drives (replacing other methods of speed control)	1980-1981	1970-1988
4. Programmable lighting controls (new technology)	1984-1985	1980-1987
5. Solid-state lighting ballast (replacing inductive ballast)	1985	1980-1987
6. Energy-monitoring and control systems (new technology)	1984-1985	1978-1989
7. Fiber optics (replacing copper conductors)	1983-1984	1980-1986
8. Multiplex fire alarm systems (replacing separately wired systems)	1983-1984	1980-1986
9. New aluminum alloys for conductors (replacing older alloys)	1982	1975-1988
10. High-technology telephone and data transmission systems (replacing older systems)	1985-1986	1982-1989
11. Amorphous-metal transformer cores (replacing laminated cores)	1989	1988-1990[a]
12. Solid-state lighting dimmers (replacing rheostats)	1983-1984	1972-1987
13. True root-mean-square meters (replacing sine-wave-only meters)	1987-1988	1985-1989

[a]One participant felt that this technology has not yet been accepted; he did not project a year of acceptance.

Note: Identified at the April 3, 1991 meeting of the Federal Construction Council Consulting Committee on Electrical Engineering; the committee used a simplified "Delphi" procedure.

ible, inherently economical, and even "democratic," because virtually anyone can use them with a minimum of special knowledge or skill.[40] A recent study of the home-building industry identified more than 100 specific innovations in housing construction that have supplemented this basic system since 1945 (see **Table E-3**); (Slaughter 1991).

Table E-3 Sample of Innovations in Permanent Residential Structures, 1945 to 1990. Source: Slaughter, (1991).

Functional Area	No. of Innovations
Structural exterior wall framing	7
Enclosure and insulation	8
Openings	13
Interior wall framing	7
Foundation	12
Floor framing	10
Roof framing	7
Roof covering	7
Plumbing	12
Electrical wiring	4
Heating/ventilation/air conditioning	12
Interior finish	18
TOTAL	117

In the areas of nonresidential building, the past several decades have witnessed the introduction of a variety of new structural materials and techniques for enclosing space and resisting loads, and for constructing these structures. **Table E-4** lists major examples.

The development of fabric tension structures can be traced to the pioneering work of Frei Otto in the 1960s, but building applications did not achieve widespread or notable commercial use until nearly two decades later with the advent of Teflon-coated fabrics, which promised longer life and better performance (Otto, 1969).

[40]This viewpoint is advanced notably in a 1978 essay by architect Andrew Rabeneck, then a principal in the San Francisco offices of The Ehrenkrantz Group.

Table E-4 Innovations in commercial structures

Tensile fabric structures
Sliding Teflon bearings
Seismic base isolation
Slurry-wall construction
Up-down construction
Fall protection on building construction
Composite steel-concrete floor construction
Metal floor and roof decks
Electrified floor construction
Single-wythe brick masonry cladding (Sarabond)
Lateral framing systems for high-rise buildings
Precast concrete construction
Tilt-up construction
Pumped concrete
High- and superhigh-strength concrete
Concrete admixtures
Concrete floor/deck hardeners
Epoxy-coated concrete reinforcing bars
Cathodic protection of rebars
Prestressed concrete
Lift-slab building construction
Staggered truss system
Pre-engineered structural systems
Tuned-mass damper for high-rise buildings (drift)
Active drift control systems for high-rise buildings
Blast-resistant (window) construction
Anti-terrorist design and construction
Single-ply membrane roofing
Curtain wall construction
Critical path method of scheduling
Ultimate strength design of concrete
Plastic design in steel
Limit state design in timber
Sprayed-on fire proofing
Weathering steel
Fire retardant ply-wood
Welded-frame system scaffolding
Motorized self-climbing scaffolding
Flying formwork
Gang-forms
Computer-aided design
Computer-aided drafting

These new fabric materials provided lightweight and relatively inexpensive cover for such large, open spaces as sports stadiums and performing arts arenas. Their development resulted from close collaboration among architects, structural engineers, and product manufacturers. The design of several very attractive buildings (notably the *Hajj* terminal in Jedda, Saudi Arabia), involving cooperation among a broad cross section of design, engineering, and product development specialists, demonstrated the new potential for coated-fabric structures.

Lightweight steel stud framing systems emerged from a combination of factors, including the desire to find a fire-resistant substitute for wood-based framing products, mainly for light commercial applications; a concerted effort on the part of U.S. steelmakers to move from automotive applications into the building industry; a general degradation in the quality and availability of dimensioned framing lumber; and participation of the U.S. gypsum industry in the development of design, engineering, and construction methods.

Design professionals, including architects, interior designers, and engineers, worked with building code officials, steel fabricators, and architectural specialty manufacturers to develop standard solutions and approaches, which continue to be developed for both residential and commercial applications.

The steel framing industry has developed a series of structural (rather than veneer, partition, or furring) applications for lightweight steel, including approaches that can be applied to low-rise multistory buildings. The brick and concrete masonry industries, which traditionally captured a larger proportion of labor and materials in such markets, have resisted these innovations.[41]

INTERIORS

Raised-floor wire management systems emerged as a direct response to the explosion in wire-based computing and communications technologies in offices and the need to provide a convenient, safe, and flexible means for handling wires. Initial raised-floor product designs, produced mainly to provide electrical continuity and adequate underfloor wire management space, failed to perform adequately from the standpoint of appearance, cost, and acoustical quality.

When architects, interior designers, and electrical engineers were retained by several key manufacturers in subsequent product design efforts, a second generation of more satisfactory raised-floor systems emerged. European product manufacturers have developed thinner raised-floor systems that do not employ structural frameworks to support removable floor tiles.

---

[41]When the brick industry developed guidelines for the use of brick veneer with steel stud backup, the engineering design provisions of these guidelines were characterized by some members of the steel framing industry as excessively conservative and intended to make steel framing approximately equal in cost to masonry.

Personal environment control furniture systems resulted from the proliferation of electronic office equipment, but they were also a response to difficulty experienced with conventional building mechanical systems in providing for human comfort in office environments. Architects, mechanical and electrical engineers, interior designers, and furniture manufacturers collaborated to develop a new concept for servicing individual workstations, based on the principle of placing controls and output devices where they are needed, rather than in remote locations.

In contrast to experience with raised-floor wire management systems, the early involvement of a broad range of design disciplines and extensive concept testing with potential users appears to have avoided unsuccessful initial results.

REFERENCES

Geller, H., J. P. Harris, M. D. Levine, and A. H. Rosenfeld. 1987. The role of federal research and development in advancing energy efficiency: A $50 billion contribution to the U.S. economy. Annual Review of Energy 12:357-397

Office of Technology Assessment. 1987. International Competition in Services: Banking Building Software Know-How. Washington, D.C.: U.S. Government Printing Office.

Otto, F., 1969. Tensile Structures: Cables, Nets and Membranes. Cambridge, Mass.: MIT Press.

Slaughter, S. E. 1991. "Rapid" innovation and the integration of components: A comparison of user and manufacturer innovations through a study of the residential construction industry. Ph.D. thesis submitted in the field of Management of Technology, Massachusetts Institute of Technology, Cambridge, Mass.

APPENDIX F

TORT LAW, DETERRENCE AND INNOVATION: TOO MUCH OR TOO LITTLE?

Michael D. Green[42]

The role that law plays in technological development is sometimes obscure but generally pervasive and fundamental. The existence of "law" as the glue that holds society together, avoiding anarchy and permitting stable and cooperative relationships to form, is a necessary condition for the modern interdependent society. Patent law, for example, gives inventors a monopoly, thereby enabling them to retain the benefits of their developments and providing financial incentives for research and innovation.

Complicating an assessment of law as technology enhancer in the construction and building context are the scope and variety of legal rules applicable to the building and construction industry in this country. In part this is due to fragmentation in the industry, which consists of a variety of commercial contractors, subcontractors, architects and engineers, residential home builders of stratified size, and manufacturers of the various materials and products used in construction. There is no such thing as "construction law," but rather bits and pieces of many diverse strands of law, which may be brought to bear on the many aspects of construction. Contract law governs the relationships among the many entities that must cooperate to produce a completed structure. Tort law plays a significant role in determining who will bear losses arising from unanticipated risks, including accidental injuries, thereby creating incentives for risk-reducing behavior. Various federal and state regulatory agencies, such as the Occupational Safety and Health Administration, have authority over pieces of the construction process.

---

[42]Professor of Law, University of Iowa

Among these several strands, tort law stands out. Highly publicized litigation and spectacular claims of risk and damage from asbestos and other building products, design and construction errors, and various manifestations of "sick building syndrome" and other operational hazards have been part of a "tort crisis"[43] that is purported to have a stifling effect on innovation in a broad spectrum of productive activity in this country. Many critics, from both the left and the right, are harsh in their assessment of tort law and would replace or restructure it in significant ways.[44]

The critique that is most relevant to issues of new technologies and innovation in the construction-related industries is that tort law has greatly increased the costs of certain activities, driven useful products off the market, stifled innovation because of fears of liability, and harmed industry's ability to compete in an international marketplace. One particular variant of this critique is that tort law is biased against high-technology, mass-produced, widely dispersed risk—so-called public risk. By contrast, private risks—those of

---

[43]The most recent crisis appeared to peak around 1985-1986, just about a decade after the crisis of the 1970s.

[44] The most strident critic on the right is Peter Huber of the Manhattan Institute, who recently published a scathing critique of developments in tort law over the past several decades. P. Huber, *Liability: The Legal Revolution and Its Consequences* (1988). Huber has condemned modern products liability law as the "courts' . . . most ignominious failure." Id. 171. Judge Alex Kozinski of the United States Court of Appeals for the Ninth Circuit, in a review of Huber's work wrote that it "is a tale of hubris and greed: hubris on the part of the Founders [the common law judges who developed modern products liability], who unhesitatingly swept aside legal rules distilled through the collected wisdom of generations of common law jurists; greed in abundance on the part of the tort lawyers who eagerly pressed the frontiers of liability outward." Kozinski, *Torts Are No Piece of Cake*, Wall St. J., Oct. 6, 1988, at 16, col. 5.

Huber, and a number of other prominent critics, favor a move toward contract-like or contract-mimicking arrangements for allocating the risk of accidental injury. Coleman, *A Market Approach to Products Liability Reform: A Theoretical Synthesis*, BNA Prod. Safety & Liab. Rep. 463 (May 5, 1989); Schwartz, *Proposals for Products Liability Reform: A Theoretical Synthesis*, 97 Yale L.J. 353 (1988). On the left, Stephen Sugarman would do away with the tort system in favor of a no-fault compensation scheme. See, e.g., Sugarman, *Doing Away with Tort Law*, 73 Calif. L. Rev. 555 (1985). Professor Jeffrey O'Connell has advocated adoption of no-fault compensation systems for the past quarter century. See, e.g., R. Keeton and J. O'Connell, *Basic Protection for the Traffic Victim* (1965).

personally controlled, old and familiar mechanisms—receive less concern, yet paradoxically are less safe. For example, the public outcry over asbestos in schools and public buildings (a public risk) has resulted in its removal, even in instances when it would be safer to leave it in place. This bias, according to the critique, stifles progress, not only depriving us of the benefit of new technology, but also leaving us with greater residual risk.[45]

This appendix considers the role that tort law plays in affecting commercial behavior. The impact of these behavioral changes on innovation, primarily as the end product of the research and development process, and the international competitiveness of domestic firms frame the inquiry. The primary focus of this discussion is the subset of tort law known as products liability, which governs the liability of manufacturers and sellers of products. This area has rapidly developed over the past quarter of a century, to become very high profile, and is likely the most significant area of tort law bearing on the question at hand. To a lesser degree, the law applicable to professional misfeasance—especially architects and engineers—may play a role in the issues of interest.[46]

The discussion begins with a brief account of the historical development of tort law, its modern goals, and its expansion over the past quarter of a century. It then proceeds to sketch out the theoretical debate over the role of tort law in regulating modern technology. In this discussion, the tensions between limiting law to advance technology and employing law to minimize risk are revealed. Next, the impact of tort law on the international competitiveness of domestic producers is considered.

---

[45]Huber, *Safety and the Second Best: The Hazards of Public Risk Management in the Courts*, 85 Colum. L. Rev. 277 (1985).

[46] For a discussion of the contexts in which architects and engineers may be liable for personal injuries, see Comment, *A Defense Catalogue for the Design Professional*, 45 UMKC L. Rev. 75 (1976).

Professional liability of architects and engineers predominantly involves claims by owners or contractors when cost overruns, delays, or construction defects appear. See INA Corporation, *Professional Liability Loss Control* A-6 (1980). Those problems, endemic in construction, generate lawsuits that determine which of the entities involved in the construction venture should bear the associated losses. Because the parties stand in a contractual relationship, thereby facilitating allocation of the risks to the party best able to control them and willing to bear them, these suits should not have much of an effect on attempts at innovation. For a discussion of legal theories and issues raised by building component failures, see Martell and Glewwe, *Building Component Failures--Sources of Salvage* in ABA, *Risks in Construction: New, Increased, Decreased* (1986).

Ultimately, two questions emerge: (1) what is the impact of tort doctrine on technological development, and (2) what effect does this impact have on net social welfare? The first question is empirical; the second is both empirical—in requiring information about the costs of innovation dampening—and a question of policy—in trading off economic development for safety. Recent work by The Conference Board and the Rand Institute provides some sketchy evidence that bears on the first question, but no systematic data exist on the benefits of tort regulation, especially in the matter of dangerous products, services, or techniques that were prevented from being placed on the market. The claim that American products liability law hinders domestic companies' international competitiveness is often stated but infrequently documented. Two distinct concerns are suggested.

One concern is that the American consuming public pays a high price, compared to consumers worldwide, for the U.S. tort system. This claim questions the trade-off between commercial productivity and safety (i.e., the basic cost-benefit question addressed earlier). Suffice it to say that it seems rational that U.S. society, given its wealth, would value human safety higher than a Third-World country anxious to accelerate economic development. As a result, products of firms marketed in the U.S. and subject to domestic liability laws may have higher prices, reflecting the additional safety required or higher liability costs.

The second concern is that domestic firms are disadvantaged by U.S. tort law in competing with foreign firms. The stringent domestic products liability system, in this view, creates an uneven playing field, in which domestic manufacturers, subject to U.S. law, must produce their products with less risk, which means added expense (to build in safety) or decreased functionality. Moreover, miscalculations are costly—manufacturers sued in the United States are subject to much higher damage awards than anywhere else in the world.[47] Thus, it comes as no surprise that American companies pay a greater amount for product liability insurance than their international competitors. The U.S.

---

[47]This disparity was recently highlighted in Dow Chemical Co. v. Alfaro, 786 S.W.2d 674 (Tex. 1990), in which Costa Rican citizens allegedly seriously injured (rendering them sterile) by chemicals produced by Shell Oil Company and Dow Chemical Company attempted to sue those corporations in Texas. An affidavit submitted by a Costa Rican judge stated that the maximum plausible recovery for the plaintiffs in Costa Rican courts was $1,080. Id. at 683 n.6. See generally McGregor, *Personal Injury and Death,* in XI *International Encyclopedia of Comparative Law: Torts* 9-169 to 9-179 (A. Tunc, ed, 1983).

Department of Commerce found that some companies pay 20 to 50 times what their foreign competitors pay for products liability coverage.[48]

On balance, precious little evidence emerges one way or the other on international competitiveness, although most of the proponents of a negative effect are long on rhetoric and short on analysis and data. Hence, this particular aspect of the debate is not treated further here.

BRIEF HISTORY OF TORT LAW

The tort system of several centuries ago would be quite unfamiliar to us today. Personal injury lawsuits played but a minor role in a society in which agriculture and land had dominant roles. Interactions between strangers were uncommon, and the engines that fueled industrialization (and wreaked their havoc on those unfortunates who got in the way) were yet to be developed.[49] Society was no better-off health-wise, indeed it was considerably worse-off, but the causes of death, illness, and injury largely resulted from natural sources rather than human-developed technology and societal interactions. Strict liability was largely the rule applied to those harms that the courts recognized as subject to liability, but that was a considerably limited class of harms. As the number of incidents causing injury grew and the scope of injuries that the courts recognized expanded, movement toward a fault-based rule of law developed. "[T]he modern negligence principle in tort law seems to have been an intellectual response to the increased number of accidents involving persons who had no preexisting relationship with one another—'stranger' cases."[50] Some historians have postulated that the movement away from strict liability in the late nineteenth century was for the purpose of aiding the development of fledgling industry.[51]

---

[48]P. McGuire, *The Impact of Product Liability* 4 (1988).

[49]The railroad, surely one of the most important technological developments in the history of mankind, is credited with playing a major role in the development of tort law in the late nineteenth century. F. Vandall, *Strict Liability: Legal and Economic Analysis* 6 (1989).

[50]G. White, *Tort Law in America* 16 (1985).

[51]The most prominent advocates of the "subsidy theory" include L. Freidman, *A History of American Law* 409-427 (1973) and M. Horwitz, *The Transformation of American Law*, 1780-1860, at 85 (1977). Gary Schwartz disputes that account in Schwartz, *Tort Law and the Economy in Nineteenth-Century America: A Reinterpretation*, 90 Yale L.J. 1717 (1981).

By the early twentieth century, fault—expressed as negligence—had captured the day. American courts resisted the British movement toward pockets of strict liability and required a plaintiff to prove negligence—a failure to exercise reasonable care—on the part of the defendant in order to recover damages. At the same time, some areas were identified for special treatment. In the workplace, occupational injuries were removed from the tort system. A no-fault compensation system was adopted that guaranteed all injured workers a modicum of recovery, thereby rendering fault irrelevant.

Throughout the twentieth century, numerous tort decisions have expanded the scope of liability, as one prominent example illustrates:

In the products liability arena, one of the most significant doctrines limited a manufacturer's obligations to those persons purchasing the product from the seller. This requirement of "privity of contract" insulated most manufacturers from suit by consumers, where intermediate distributors existed. The privity requirement was swept away in a seminal decision by Judge Cardozo in *MacPherson v. Buick Motor Co.*,[52] in 1916, greatly expanding the potential liability of product manufacturers. With *MacPherson*, the framework for the development of modern products liability law began. Henceforth, manufacturers would owe a duty of due care to all who might foreseeably be harmed by the use of the product.

Despite the signal revolution in liability law marked by *MacPherson*, the privity requirement remained in place for many nonproduct claims. Thus, architects and engineers were not liable to third parties with whom they had not contracted. Not until 1957, when the New York Court of Appeals[53] analogized a building to a product and held that architects could be sued by third parties, was the privity barrier eliminated for architects and engineers. Most other states have followed suit.

Tort law has grown in tandem with the industrialization of society and technological development. Industrialization and technology have improved society's well-being immeasurably, including health and life expectancy, but have at the same time transferred many risks from natural causes to human-controlled causes. Natural risks were of little interest to the tort system, but transferring risks to human control creates numerous opportunities for human fallibility, error, and harm. Kidney dialysis machines, for example, prolonged the lives of many with kidney disease, but at the same time created opportunities

---

[52]217 N.Y. 382, 111 N.E. 1050 (1916).

[53]Inman v. Binghamton Housing Authority, 3 N.Y.2d 137, 143 N.E.2d 895, 164 N.Y.S.2d 699 (1957).

for human error in the operation of this technology that could cause significant harm.[54] Thus, the shift of risks to human control has fueled significant growth in the tort system.

Probably the most significant factor in the development of modern liability law is strict products liability and its theoretical enterprise liability underpinnings (see next section for definition of enterprise liability). The intellectual lineage of strict products liability is quite extensive, but one figure who stands out is a judge on the California Supreme Court, Roger Traynor, who in 1944 began a crusade for imposing strict liability on manufacturers.[55] Twenty years later he persuaded a majority of his colleagues on the California Supreme Court to impose strict liability on the manufacturer of a Shopsmith, which could be used as a variety of power tools. While the purchaser's husband was using the Shopsmith as a lathe, set screws that were inadequate to hold the wood stock in place gave way and the wood flew out of the machine, hit the plaintiff in the head, and caused serious injuries.

Rejecting both sales law, an area of contract law, and fault-based tort notions, which the defendant argued should govern its liability, Justice Traynor wrote:

> A manufacturer is strictly liable in tort when an article he places on the market, knowing that it is to be used without inspection for defects, proves to have a defect that causes injury to a human being. . . .
>
> The purpose of such liability is to insure that the costs of injuries resulting from defective products are borne by the manufacturers that put such products on the market rather than by the injured persons who are powerless to protect themselves.[56]

Over the remainder of the decade, the move toward strict products liability would receive widespread support and adoption in the supreme courts of most states.[57] The American Law Institute, an influential organization of attorneys,

---

[54]Grady, *Why are People Negligent? Technology, Nondurable Precautions, and the Medical Malpractice Explosion*, 82 Nw. U.L. Rev. 293 (1988).

[55]Escola v. Coca-Cola Bottling Co., 24 Cal. 2d 453, 150 P.2d 436 (1944) (Traynor, J., concurring).

[56]Greenman v. Yuba Power Products, Inc., 59 Cal. 2d 57, 27 Cal. Rptr. 697, 377 P.2d 897 (1963).

[57]As with virtually all tort law, products liability is a creature of state common law. This means that each state has power to shape its own law, creating the possibility of variations and inconsistency. It also means that

judges, and academics, published a strict liability provision that influenced many state courts.[58] During this period, many jurisdictions expanded the areas to which this new strict liability applied, including leased products, used products, and completed buildings.[59]

In parallel with these technical legal developments, there have been shifts in public as well as judicial attitudes about the acceptability of risk, control of risk, and responsibility for risk, which help explain at least some of the expansion of liability and the upward movement in tort damages over the past several decades. This movement, difficult to document comprehensively in precise terms, is exemplified by the $125 million jury verdict for punitive damages in the Ford Pinto case in 1978 in California.[60] The $125 million jury award was reduced by the court to $3.8 million.

PURPOSE OF THE TORT SYSTEM[61]

The twin goals of the tort system should be to minimize the costs of accidents and to effectuate fairness in allocating the losses due to accidents. Because the concept of fairness is elusive and variable, it might be viewed as a veto or constraint on the more analytical, and ultimately utilitarian consideration of minimizing the costs of accidents. Tort law should provide incentives to minimize the sum of the costs of accidents *and* the costs of accident prevention. This view—termed the enterprise theory of tort law—acknowledges that some accidents are not worth avoiding.

The costs of accidents fall into three primary categories. The first category is direct costs and consists of two components: (1) the costs of preventing or avoiding those accidents for which it is cost-effective to do so; and (2) the losses suffered by victims of the remaining accidents—the loss of wages, costs of medical care, and less quantifiable costs of losing a limb or the ability to engage in desirable activities, including pain and suffering. It is important to keep in

---

courts, rather than legislatures, are the primary institution making this law.

[58]Restatement (Second) Torts § 402A (1965).

[59]See, e.g., Schipper v. Levitt & Sons, 44 N.J. 70, 207 A.2d 314 (1965) (holding home builder liable for defect in a water heater installed in home); see generally Comment, *Strict Liability and the Building Industry,* 33 Emory L.J. 175 (1984).

[60]Grimshaw v. Ford Motor Co., 119 Cal. App. 3d 757, 174 Cal. Rptr. 348 (1981).

[61]Much of this section is derived from an influential book entitled *The Costs of Accident*, authored by Dean Guido Calabresi of the Yale Law School.

mind that the latter component of cost is borne by society regardless of whether the victim is compensated. Direct costs do not disappear when a defendant is not required to pay for them, although they do remain more hidden and outside such contemporary measures of welfare as the Gross National Product.

The second category of costs requires recognition of interpersonal comparisons of utility—a catastrophic loss to one individual is a greater loss than if it is spread among a larger group. The prevalence of insurance in contemporary society is powerful evidence of the reality of these costs, but whether government should impose this view through law by building an insurance component into strict products liability law or leave such judgments to the marketplace is a more controversial matter.

Finally, a third category of the costs of accidents is the money society spends to decide whether to reallocate accident costs from the victim to someone else—friction costs of the tort and litigation system. The costs of paying lawyers and expert witnesses and maintaining a court system are as surely a cost of accidents as are the auto body repairmen's bills.

The first step in effectuating the cost minimization goal is to provide incentives to take precautions against accidents for which the costs of avoidance are less than the costs of the accident. This is the "deterrent" function of tort law. Imposing liability for accidents on each activity or product that causes them should effectuate this goal by requiring the entity to internalize the costs of accidents. This internalization creates the deterrence that the tort system in theory provides. Imposing liability on the entities responsible for activities posing risks provides the financial incentives for cost-justified safety precautions. The responsible entity, then, should invest in safety until the next dollar spent will reduce accident costs by something just less than that dollar.

Of course, the injured victim is one of the entities who caused the accident and upon whom the costs might be imposed to effectuate deterrence. Imposing liability on the victim, however, rarely has a significant deterrent effect—most of us drive carefully or otherwise take precaution not because we are concerned about the impact of our behavior on some future lawsuit, but because we wish to avoid the unpleasant prospect of pain, injury, or death. By contrast, producers, as artificial entities concerned with profit maximization, are far more susceptible to behavior modification through financial incentives provided by the legal system.

As a result of this internalization of accident costs, the price of a product or activity will generally reflect the costs of those accidents for which avoidance is not worthwhile (i.e., is too costly). The higher price will cover that residual level of risk and accidents that cannot be reasonably avoided. For these residual accidents, the direct costs of the accident may be reduced by providing prompt compensation to the victim to mitigate the damages suffered. Thus, prompt compensation may enable an accident victim to obtain a prosthesis or occupational rehabilitation, which will reduce the extent of disability suffered.[62] Even if no direct savings can be obtained, payment to the victim serves the loss-spreading function, thereby reducing secondary costs.

This compensatory aspect of tort law has been the predominant influence in tort law's modern developments. Courts have focused on finding a source, sometimes quite distant from the locus of responsibility, to provide compensation to injured plaintiffs. Yet this emphasis on compensation tends to obscure the impact on deterrence, to which it is only loosely tied, and deterrence often receives secondary consideration from the courts.[63] In particular, the amount of compensation that is optimal may not be the same amount the party held liable should pay to optimize deterrence.[64]

The expansion of liability over the past several decades has been supported by this enterprise theory of tort law. Although the contemporary tort system is not modeled perfectly on these economic underpinnings, design defects, by far the largest and most significant class of products liability cases, are generally determined by reference to a cost-benefit analysis of the challenged design.[65] Under this scheme, manufacturers have incentives to take all cost-justified precautions, but the costs of those accidents not worth avoiding are left with the victims, rather than imposed on manufacturers.

DETERRENCE: TOO MUCH OR TOO LITTLE

Proponents argue that if the tort system, viewed within this economic framework, is operating optimally, societal resources are maximized. Those risks worth avoiding will be avoided by manufacturers acting in their own economic self-interest; those risks not worth ameliorating or avoiding will remain. Many contemporary commentators have argued that the tort system is not operating optimally. In particular, it is claimed that current operation of the system is

---

[62]See statement of Leonard Bender, M.D., on behalf of the American Congress of Rehabilitation and the American Academy of Physical Medicine and Rehabilitation, in Hearings before the Subcommittee on Consumer Protection and Finance of the House Committee on Interstate and Foreign Commerce, 95th Cong., 1st Sess. 594 (1977).

[63]The hegemony of compensation in modern tort law might be blunted if universal health-care coverage were available. In other Western countries with universal healthcare, such as Canada and England, compensation does not play the same influential role as it does in this country.

[64]See S. Shavell, *Economic Analysis of Tort Law* § 10.2 (1987).

[65]See M. Shapo, *The Law of Products Liability* 9-9 to 9-11 (1987).

excessive in its deterrence, leading manufacturers and other producers to determine that costs of new products are too great to merit their introduction to the marketplace.

Before turning to the possible reasons for overdeterrence, a note of caution is in order. Much of the literature documenting, discussing, and analyzing the 1980s tort crisis has focused on the rate of change that occurred rather than on whether those changes are making the system more or less efficient.[66] Thus, statistics documenting the increase in the number of tort suits filed or the increase in the median damage award do not tell us whether the system was underperforming in the past or is overdeterring currently. Similarly, reports of the increased impact on corporate operations, rapid increases in liability insurance premiums, and other responses to expanded liability do not answer the question of whether those occurrences are steps toward or away from optimization of the system. Although rapid changes no doubt cause dislocations, and discomfort, and unsettle planning, the elusive ultimate question is not the rate of change of the system, but the direction in which it is moving.

Among the plausible reasons why overdeterrence may occur is error in the judicial process. Courts may make errors in performing cost-benefit analyses, systematically favoring plaintiffs. Juries are allocated substantial power in deciding tort cases, and most observers are inclined to believe that jurors sympathize with badly injured plaintiffs. The "public risk" version of this proplaintiff bias hypothesizes that if jurors (and even scientifically untrained judges) react irrationally to high-technology, high-profile, public risks (such as nuclear power plants or toxic waste dumps), then excessive liability may be imposed on producers of what logic and scientific measurement show to be safer technology.[67]

---

[66]See, e.g., Malott, *Product Liability System Hampers Effectiveness*, Financier, Jan. 1988, at 29 (arguing that growth of number, size, and uncertainty of products liability awards has caused increased prices, led to withdrawal of products from the market, and hampered innovation).

[67]See, e.g., *Counting on Science at EPA*, 249 Science 616 (1990) (detailing the disparities between a recent Environmental Protection Agency ranking of the most significant health and ecological risks and the public's ranking of those concerns).

Professors Clayton Gillette and James Krier make a powerful argument that simply counting fatalities as most expert risk assessors do, misses a different, yet rational world view to which the public subscribes: an aversion to involuntarily imposed (as opposed to voluntarily assumed) risk; dislike for risks involving latency periods and insidious disease as opposed to sudden traumatic injury; and intolerance for man-made as opposed to natural risks. Those preferences are not captured in standard mortality and morbidity statistics. Gillette

Thus, the public fears airline travel more than automobile travel, yet, in terms of mortality and morbidity, airplanes are considerably safer than cars. This bias might take the form of imposing liability incorrectly—false positives—or of being overly generous in awarding damages. The latter bias is possible because of the vague standards governing awards of nonpecuniary damages, thereby leaving substantial discretion with juries as to the appropriate amount of the loss.[68] If either of these phenomena occurs, producers should respond either by building more safety into their products than is sound, by raising the price of their goods to reflect a premium for judicial error, or by simply refusing to develop or market these technologies and services. The last of these responses is most likely to occur when substantial uncertainty pervades either future judicial treatment of new goods and services or the extent of risk contained in them.[69]

A second potential source of excessive deterrence is the availability of punitive damages, which have become much more prevalent and prominent in the past 15 years. These damages, which do not reflect any losses suffered by the plaintiff, could result in imposing excess liability on producers. Moreover, because of the vague standards regulating the appropriate amount of such damages, the uncertainty thereby created may result in excessive precaution by the risk-averse manufacturer.

A third possible source of overdeterrence stems from uncertainty, both factual and legal. Legal uncertainty exists because of the expost, case-by-case decision making employed by tort law and the imprecise standards that permit fact finders substantial freedom in deciding liability questions and setting the amount of damages. Jury decisions as to whether a product is defective occur many years after manufacturers complete the design process. Moreover, each jury decides liability unencumbered by previous jury determinations about the same product or conduct, which sometimes results in inconsistent decisions. No precise safety standards exist to provide a safe harbor to producers. Compliance with federal or state regulatory standards does not generally preclude tort liability. Moreover courts are faced with a particularly complex problem when

---

and Krier, *Risks, Courts, and Agencies*, 138 U. Pa. L. Rev. 1027 (1990).

[68]Juries are assigned the role of assessing the damages suffered by victims and determining an appropriate amount to be paid to the plaintiff to making her whole. A necessary by-product of that determination, because of the inextricable linkage between plaintiff and defendant, is that it also values the cost of the accident, providing information to defendants about how much should be spent on accident avoidance.

[69]Calfee and Craswell, *Some Effects of Uncertainty on Compliance with Legal Standards*, 70 Va. L. Rev. 965 (1984).

required to determine retrospectively how much risk a manufacturer anticipated and whether the extent of investigation into safety at the time was appropriate in light of what was known.[70] Consequently, producers must guess how, in the aggregate, juries will respond in the future when and if their product causes injury or disease.

In addition, factual uncertainty exists because producers can never be sure about the precise amount of risk in a new product or new technology, and obtaining better information is costly. New technology, in particular, is most likely to have unknown elements that will result in unanticipated failures.[71] Numerous examples of unanticipated dangers that were only revealed when the new technology was put into operation include the failure of turbine blades in the Boeing 747 and the unforeseen resonances that weakened engine mounts in the Electra, resulting in fatal crashes.[72]

Some have argued that the tort system underdeters. For example, access barriers may prevent a valid claim from being asserted. Probably the most significant such barrier is the difficulty and expense of gathering and presenting evidence to establish a plaintiff's claim. Since the burden of proof is routinely placed on the claimant, unavailability of evidence may make a lawsuit infeasible or reduce its settlement value. Moreover, before a claim can be asserted and pursued an injured victim must recognize that there is legal redress available. These biases tend to work systematically in favor of defendants and against plaintiffs. They are likely to be most pronounced in high-technology areas, such as environmental, nuclear, or toxic substances, where substantial uncertainty exists about causal mechanisms. It may also be substantial in long-latency disease cases, where the passage of time may result in the deterioration of evidence.

Another form of access barrier occurs when the harm suffered is minor, especially when many small harms occur that are spread over a large and diffuse group of people, such that it is not economically feasible to pursue private damage claims because of the costs of legal action.[73] Statutes of limitations pose yet another access barrier. Although these statutes contribute to efficiency in litigation, they exact a toll of false negatives, both for cases that are filed too late and for incipient claims that are never filed because plaintiffs (or their

---

[70]See S. Shavell, *Economic Analysis of Accident Law* 55-56 (1987).

[71]H. Petroski, *To Engineer is Human: The Role of Failure in Engineering Design* 219 (1985).

[72]N. Rosenberg, *Inside the Black Box: Technology and Economics* (1982).

[73]See K. Viscusi, *Regulating Consumer Product Safety* 14 (1984): "If a consumer is injured in a minor way . . . a product liability lawsuit will not be financially attractive."

lawyers) recognize the futility of filing a claim in the face of an expired statute of limitations.

Other possible sources of underdeterrence include agency cost problems in which corporate managers do not act in conformity with the best interests of the corporation but rather tend to favor their own personal interests. This effect may also occur most prominently for risk with long time lines. Corporate executives or officials, under pressure to perform in the short run and unlikely to still be in a responsible position decades later when the risk manifests itself may not take precautions that would otherwise be optimal for the organization and for society.[74] Similarly, the limited liability of corporations could result in excessive risk taking. Low-probability, high-magnitude risks may be financially attractive for a firm that will not bear all of the costs if the risk comes to fruition,[75] although this possibility is rendered less likely because of management's desire to retain its position and control of the corporation, at least for short-run risks.[76]

IMPACT OF THE TORT SYSTEM

What do the data show about the impact of tort law on effecting an optimal level of deterrence? Does tort law affect producer behavior? If so, is there an impact on innovation? A few recent studies have investigated the effect of products liability law on corporate behavior. These studies have tended to focus on the impact of *change* in product liability laws, assessing the consequences of expanded liability. However, assessing responses to change does not resolve the overall question of effect, which remains open. One recurring theme that emerges from all of the studies is the industry-specific nature of the impact. What one might say about the small aircraft industry, which has been heavily affected by strict product liability, surely could not be said of the steel industry. No study focuses on or considers what might loosely be termed the construction industry.

---

[74]See Felstiner and Seigelman, *Neoclassical Difficulties: Tort Deterrence for Latent Injuries*, 11 Law & Pol'y 309, 309-912 (1989); Hayes and Abernathy, *Managing Our Way to Economic Decline*, Harv. Bus. Rev., July-Aug. 1980, at 67; Henderson, *Product Liability and the Passage of Time: The Imprisonment of Corporate Rationality*, 58 N.Y.U.L. Rev. 765 (1983).

[75]See Shavell, *The Judgment Proof Problem*, 6 Int'l Rev. L. & Econ. 45 (1986).

[76]Roe, *Corporate Strategic Reaction to Mass Tort*, 72 Va. L. Rev. (1986).

An early 1980s study performed by the Rand Institute for Civil Justice investigated how firms had responded to the development of strict product liability. Interviews were conducted with corporate product safety officers of nine large consumer products corporations that are considered leaders in the safety field. The study revealed that among several forces that have combined to focus attention on product safety, "product liability is the most significant influence on product safety efforts." The authors distinguished industries based on how heavily regulated they were, concluding that in industries subject to less governmental safety regulation, "products liability probably exerts the overwhelming pressure" influencing design decisions.[77] Only in industries subject to heavy regulation (the airline and pharmaceutical industries, for example) does regulation outweigh products liability law regarding safety-related decisions. However, the question of disincentives to innovation was not considered.

Two studies conducted by the Conference Board, a nonprofit corporation that provides information to business executives, did consider the impact of liability laws on technological innovation. The first study,[78] conducted in the midst of the 1980s torts crisis focused on:

> what, in fact, has been the impact of the liability and/or insurance crisis on the nation's major corporations and, by extension, on the economy? . . . How in fact has the corporate world reacted to both current insurance prices and to lawsuits filed by aggrieved consumers?

The author surveyed 232 risk managers in major domestic manufacturing, trade, and service corporations. The study was limited to large corporations; only those with a minimum of $100 million annual sales were included. The most surprising conclusion of the study was the insignificance of liability on corporate operations:

> The most striking finding is that the impact of the liability issue seems far more related to rhetoric than to reality. Given all the media coverage and heated accusations, the so-called twin crises in product liability and insurance availability have left a relatively minor dent on the economics and organization of individual large firms, or on big business as a whole. In the words of one manager: "There may be less here than meets the eye."

---

[77]G. Eads and P. Reuter, *Designing Safer Products* 122 (1983).
[78]N. Weber, *Product Liability: The Corporate Response* II (1987).

> Where product liability has had a notable impact—where it has most significantly affected management decision making—has been in the quality of the products themselves. Managers say products have become safer, manufacturing procedures have been improved and labels and use instructions have become more explicit.

The study found substantial differential impact of liability on different industries. Two-thirds of respondents reported that the cost of liability insurance (which includes only a portion of liability costs) contributed no more than 1 percent of the price of their goods. Another 11 percent reported that final prices increased by 2-3 percent. However, in certain high-risk industries—sporting goods, for example—25 percent of the price of some products was attributable to liability costs.

The impact on innovation was addressed by asking respondents whether they had declined to market a new product or service: 13 percent of the responding firms indicated that they had made such a decision in response to liability concerns, providing some solid evidence that products liability does affect the level of innovation, although apparently only in a small percentage of large firms.

The Conference Board undertook a second study of the impact of products liability by surveying the chief executive officers (CEOs) of 2,000 large manufacturing firms as well as CEOs of 2,000 manufacturers having fewer than 500 employees. The response rate to each survey was less than 15 percent, which raises concerns about selection bias. In general, the second Conference Board study found greater concern with the product liability system, a higher degree of impact on corporate operations, and primary concern about the uncertainty of ex post liability determinations with the attendant unpredictability in the amount of damages awarded by juries. A much smaller, but methodologically similar, study of Pennsylvania corporations yielded data very similar to the second Conference Board study.[79] In general, the firms surveyed in the Pennsylvania study were smaller firms than in the Conference Board studies; more than 80 percent of respondents had sales of less than $50 million. Most significantly, the second Conference Board study found a far more profound impact on innovation. Of the large firms surveyed, 24 percent of the CEOs reported that their firms had decided against introducing new products, more

---

[79]Linneman and Ingberman, *Product Liability Law: The Economic Impact on Pennsylvania* 6 (unpublished paper).

than twice the rate found in the earlier study.[80] Similarly, 23 percent of the respondents from small firms had made the same decision.[81] Also, 16 percent of the large firms reported that they had discontinued product research, whereas 13 percent of the small firms had done so.

Of considerable concern among companies in the durable goods industry was the liability "overhang" for products that had been produced years or decades ago when liability standards were not as strict. Manufacturers are subject to contemporary standards when a product manufactured years ago fails and causes harm. This creates a competitive advantage for new entrants into the field, who are not saddled with this liability overhang.

The survey of CEOs also found a greater impact on costs than the first Conference Board study: 38 percent reported a "major" impact of products liability on direct costs, which was not explicitly defined, but appears to be in the 10-15 percent of total costs range and higher.[82] Only 18 percent of respondents reported a 2 percent or less increase in direct costs. While the "direct cost" category in the second Conference Board study was broader than the liability insurance costs studies in the first survey (direct costs included litigation costs and payments made under deductibles), the different definitions are unlikely to account fully for the disparities. The second survey did confirm the earlier finding of disparate impact depending on the industry involved.[83]

A study conducted by the American Textile Machinery Manufacturers' Association[84] found that 12 percent of surveyed machinery manufacturers

---

[80]The 24 percent figure is based on a reanalysis of the data reported, which was broken into two different groups based on whether the reporting firm had actual liability experience or was acting on anticipated liability concerns. The author of the study erroneously added the percentage reported by each of the groups when giving the composite figure.

[81]Of the respondents in the Pennsylvania survey, 25 percent reported deciding against new products. However, only 2 percent of the Pennsylvania corporations reported discontinued research.

[82]A similar figure, 34 percent, emerged from the survey of smaller manufacturing firms located in Pennsylvania.

[83]The differential impact was also found in the survey of Pennsylvania corporations: "We found that certain types of firms are much more strongly impacted by product liability concerns than others." Linneman and Ingberman, *Product Liability Law: The Economic Impact on Pennsylvania* 6 (unpublished paper).

[84]American Textile Machinery Manufacturers' Association, *International Study of Product Liability Costs and System for Five Domestic Machinery Industries* (1984).

reported having decided against the development of a new product line. The reported lower figure by machinery manufacturers (compared to the other studies) may be a consequence of the susceptibility of machinery to be designed for additional safety by providing more safeguards. By contrast, chemicals, drugs, and insulation materials often cannot be made safer and must be marketed as formulated or withheld from the market.

From these studies, it seems fair to conclude that the expansion of liability has had a measurable and not insignificant effect on innovation. The extent of that effect is difficult to measure precisely, and likely varies across industries. High-technology industries, in which mechanisms of risk are less well understood and products cannot be altered incrementally to improve their safety, are subject to a greater impact than other industries.[85]

Recognizing a measurable impact, however, does not answer the question of whether that effect is socially undesirable. The missing pieces are the safety gained by the suppressed innovation and the marginal benefit of the suppressed innovation over existing technology. If the products withheld were the next generation of thalidomide, society has plainly benefited from the deterrent effect identified. If, on the other hand, producers are erring in their assessments of the risks posed by overestimating the risks, or because of uncertainty are acting in a risk-averse fashion in deciding whether to develop and market new technology, or even responding accurately to a systemic bias in favor of injured plaintiffs, then social welfare is diminished. Unfortunately no data exist to assess the costs and benefits of withheld products or lost technology. A Rand Institute study of the effect of products liability on design decisions concluded: "Data do not exist to permit judgment of the reasonableness of the current system. It is not possible to measure the improvement, if any, in the level of safety of consumer goods that has resulted from changes in regulation and law."[86]

However, there can be little doubt that significant safety benefits have been realized from the liability system. The first Conference Board study emphasized that effect, and a recent Rand Institute study observed the following:

> One of the clear implications of the information we have collected, however, is that firms have changed their behaviors in some ways that are consonant with the goals of those who advocate expanded liabili-

---

[85]See, e.g., Broad, *Does the Fear of Litigation Dampen the Drive to Innovate*, N.Y. Times, May 12, 1987, § C, at 1, col. 1 (reporting on decision to forgo development of a powerful particle accelerator to irradiate food because of unknown risks that might exist in the process).

[86]G. Eads and P. Reuter, *Designing Safer Products* (1983).

ty. . . . [T]he growth of product liability awards and of strict liability has increased the sensitivity of many corporations to product safety.[87]

TORT SYSTEM TRANSACTION COSTS

One aspect of costs for which good data do exist is the transaction costs of the tort system. If the amount actually received by injured claimants is said to represent the costs of accidents for which avoidance is not worthwhile, transaction costs represent the costs of administering those transfer payments. These transaction costs are to some extent inevitable, but different administrative systems result in different costs. For example, a major benefit of no-fault insurance is that it eliminates the costs of determining whether an individual was negligent, thereby reducing the proportion of transaction costs.

The most recent data on the total costs of the tort system, broken down into accident and transaction components, are set forth in **Table F-1**.

As Table F-1 shows, net compensation to injured victims is only 46 or 47 percent of the total tort system costs. Put another way, this means that of every dollar paid by defendants for insurance, self-insurance, and opportunity costs for efforts involving litigation, only 46 cents is received by injured persons. The 46 cents per dollar figure is an average—the efficiency of the tort system varies depending on the type of accident—and when automobile-related claims (which have become quite routinized) are excluded, the net compensation received by plaintiffs drops to 43 cents per dollar. A Rand Institute study of costs in asbestos litigation found that plaintiffs received 37 cents of every dollar expended.[88]

GOVERNMENT CONTRACTOR DEFENSE[89]

Proponents of the view that the tort system excessively deters innovation suggest that government's limited immunity from tort suits offers an important opportunity. Government should undertake, so the argument goes, to be the innovator, providing a sheltering umbrella to protect in the process from tort liability that may hinder innovation.

87P. Reuter, *The Economic Consequence of Corporate Liability: An Exploratory Study* 3 (1988).

88J. Kakalik, P. Ebener, W. Felstiner and M. Shanley, *Costs of Asbestos Litigation* (1983).

89Much of this section is adapted from Green and Matasar, *The Supreme Court and the Products Liability Crisis: Lessons from* Boyle's *Government Contractor Defense*, 63 S. Cal. L. Rev. 639 (1990).

TABLE F-1 Tort Systems Costs—State and Federal Courts, 1985.

	Dollar Value (billions)	Percentage of Total
Public expense (court system)	.5	1-2
Plaintiffs' legal fees and expenses	6.3-7.6	21-22
Plaintiffs' time and related expense	0.7-1.1	2-4
Total plaintiffs' expense	7.0-8.7	24
Defendants' legal fees	4.7-5.7	16
Insurance company costs	0.8	2-3
Defendants' time and related expenses	2.5-3.5	9-10
Total defendants'expenses	8.0-10.0	27-28
Compensation received by plaintiffs	13.7-16.4	46-47
Total costs	29.2-35.6	

The argument gains weight and currency from the 1988 Supreme Court decision in *Boyle v. United Technologies Corp.*[90] On April 27, 1983, David Boyle, a military helicopter copilot, died in a crash of a helicopter manufactured by the Sikorsky Division of the United Technologies Corporation. Although Boyle survived the crash impact, he drowned when he could not escape from the helicopter before it sank in the ocean.

Boyle's estate brought suit against United Technologies, alleging that Sikorsky had defectively designed the copilot's emergency escape system: the escape hatch opened out instead of in, and thus was ineffective when submerged, and access to the escape hatch handle was obstructed. The jury returned a substantial verdict in favor of Boyle's estate, which was overturned by the United States Court of Appeals for the Fourth Circuit.

Before *Boyle* a number of lower federal courts had created a "military contractor defense" that relieved suppliers of goods to the military from liability under state tort laws. These decisions employed federal law to "preempt" the application of state law, in effect providing immunity for government contractors where the defense applied. Those courts and the Fourth Circuit relied in large

[90]108 S. Ct. 2510 (1988).

measure on the doctrine established in *Feres v. United States*,[91] which immunizes the government from liability to members of the military for injuries received in the course of their military employment. Although the *Feres* doctrine was inapplicable to the *Boyle* case because the plaintiff sued a private contractor rather than the federal government, the Fourth Circuit nevertheless found that the policies underlying *Feres* were also applicable in a suit against a contractor supplying goods to the military.

The United States Supreme Court affirmed the decision, but on different grounds. The Court rejected reliance on *Feres*, instead invoking another exception to government liability known as the discretionary function exception.[92] The essential purpose of the discretionary function exception is to foster uninhibited political, economic, or social policymaking in the executive or legislative branch by insulating those decisions from judicial review. Although, as with *Feres*, the discretionary function exception provides immunity only for the federal government, the Supreme Court believed that its concerns were also implicated when a military contractor was sued:

> We think that the selection of the appropriate design for military equipment to be used by our Armed Forces is assuredly a discretionary function. . . . It often involves not merely engineering analysis but judgment as to the balancing of many technical, military, and even social considerations, including specifically the trade-off between greater safety and greater combat effectiveness. . . . [W]e are . . . of the view that permitting "second guessing" of these judgments . . . through state tort suits against contractors would produce the . . . effect sought to be avoided . . . ,[93]

The Court then announced the elements of its new defense:

> Liability for design defects in military equipment cannot be imposed, pursuant to state law, when (1) the United States approved reasonably precise specifications; (2) the equipment conformed to those

[91]340 U.S. 135 (1950).

[92]The discretionary function exception is contained in the Federal Tort Claims Act, 28 U.S.C. § 2671 et seq. (1982), which generally makes the federal government liable for its torts, subject to a number of exceptions. The discretionary function exception, § 2680(a), excepts from liability "any claim . . . based upon the exercise or performance or the failure to exercise or perform a discretionary function or duty on the part of a federal agency or an employee of the Government, whether or not the discretion involved be abused."

[93]108 S.Ct. at 2117-2118.

specifications; and (3) the supplier warned the United States about the dangers in the use of the equipment that were known to the supplier but not to the United States.[94]

The potential application of the government contractor defense created in *Boyle* to federal construction projects is quite evident. Moreover, *Boyle* may provide a significant safe harbor for trying out innovations that would otherwise not be employed because of liability concerns, if such exist. The safe harbor would extend to suits by third parties who might suffer injury but would not encompass losses due to flaws in the building, causing loss to the government. The latter losses could be allocated however the government and its contractor decide in the construction contract.

However, it is not clearly established that the *Boyle* decision may be applied more broadly than military procurement, and two post-*Boyle* decisions reached the conclusion that *Boyle* is so limited. In *Nielsen v. George Diamond Vogel Paint Co.*,[95] a civilian painter employed by the U.S. Army Corps of Engineers to paint a dam sued the manufacturer of the paint he used, claiming that inadequate warnings supplied with the paint led to his suffering personal injuries. The court concluded that despite the *Boyle* Court's switch from *Feres* to the discretionary function exception as the basis for the government contractor defense, "the policy behind the defense remains rooted in considerations peculiar to the military."[96] Accordingly, the court held that government contractor immunity was unavailable to the manufacturer of the paint, which was produced for use in a civilian, as opposed to a military, project. An earlier decision by a federal district court judge in Hawaii had similarly concluded:

> The federal interest in *Boyle* was the procurement of military equipment to be used by the armed forces. It is clear the *Boyle* opinion applies only to military equipment. Asbestos insulation products are not military equipment. The procurement of supplies by the United States is not enough to immunize the manufacturer under *Boyle*."[97]

Despite these cases, the *Boyle* opinion and its reliance on the discretionary function exception may have broader application. The discretionary function exception is applicable to all aspects of governmental activity and has been

---

[94]Boyle v. United Technologies Corp., 108 S.Ct. 2510, 2518 (1988).

[95]892 F.2d 1450 (9th Cir. 1990).

[96]Id. at 1455.

[97]In re Hawaii Federal Asbestos Cases, 715 F. Supp. 298, 300 (D. Hawaii 1988).

regularly applied in the civilian sector. Nowhere in the *Boyle* opinion does the Court explain why military procurement is different from civilian procurement with respect to the exercise of policymaking judgment. Nothing in the Court's opinion suggests why the discretionary function rationale would be categorically inapplicable to civilian procurement.[98] Indeed, some decisions relating to military procurement may not be sufficiently policy-related to justify protection of the contractor with immunity.[99] The inquiry in all cases should be whether the decision was based on trade-offs that require a policy judgment, regardless of whether this is made in a civilian or military context. Recently a federal court adopted this reasoning and concluded that the government contractor defense is applicable to civilian procurement. In re Chateaugay Corporation, CCH Prod. Liab. Rep. ¶ 13,042 (U.S. Bankr. Ct. S.D.N.Y. Oct. 1991).

Another open question after *Boyle* is whether the contractor defense extends to defects other than those of design. Thus, if, as in *Nielsen*, the plaintiff alleges that inadequate warnings were supplied, would the claim be barred by the *Boyle* decision? The *Boyle* opinion carefully limits its language to design defects, but once again the Court's reliance on the discretionary function exception as the conceptual foundation for its decision suggests a broader exception. The key is not the type of defect, but the extent to which the government confronted safety, functionality, and innovation concerns, and approved proceeding in a fashion thought to maximize public policy.

Hence, a decision by a branch of the armed services that fostering innovation in construction is a matter of substantial importance, such that it adopts recent construction innovations and mandates that they be employed, seems likely to fall squarely within the *Boyle* contractor defense. To be sure, the governmental officials in charge of the project should document their exercise of discretion to employ new technology, and the contractor should be forthcoming in sharing its knowledge about the risks and benefits of the innovation to be employed. Information about alternatives that might be pursued within the umbrella of the innovation should also be shared. However, if these conditions are met, immunity should follow. Similar confidence cannot be expressed about the availability of immunity where the government agency is a

---

[98]The Court did state that the selection of appropriate designs for military equipment falls within the ambit of the discretionary function exception. Regardless of the truth of that proposition, it does not explain why the selection of designs for other equipment purchased by the government would not also fall within the discretionary function exception.

[99]Thus, a mindless rubber stamping by the government of detailed plans prepared by the contractor would no more fulfill the mandate of the discretionary function exception than would delegating complete design authority to the supplier.

civilian one. The reasoning of the Supreme Court in *Boyle* would support immunity, but to date the post-*Boyle* lower courts have reached conflicting decisions on this issue.

APPENDIX G

POINTS OF ENTRY FOR BUILDING INNOVATION

Facilities are developed and used in a multistage process, over a period of years. The process is often described in terms of four principal stages: (1) planning, (2) design, (3) construction, and (4) operations and maintenance, but in fact includes many more detailed and distinct steps. In the absence of any major delays, the first three stages are typically accomplished over 12 to 24 months. Operations and maintenance continue throughout the facility's several decades of service life. A fifth stage—renewal, reuse, or demolition—is beyond the scope of this discussion, but can be an effective return to the first stage of the process for many facilities.

Opportunities for innovation, for adoption of new building technology, can occur at many points in this process, through the actions of any of the large numbers of people and organizations involved. The owner, designer, and builder comprise the major participants, but each of these three is in fact a complex group of individuals and organizations that must work together to accomplish the aim of a completed project.

The initiator and central figure in the process is the facility owner (see **Figure G-1**), who is responsible for stating his or her needs and employing appropriate resources to meet those needs. The user is normally the basis for determining needs, but in many cases the owner and user are different and possibly separate organizations. In government, agencies or parts of agencies that will occupy a building may have little involvement in facility design and construction. The General Services Administration, the U.S. Army Corps of Engineers, and the Naval Facilities Engineering Command undertake facility development for other agencies as well as for themselves.

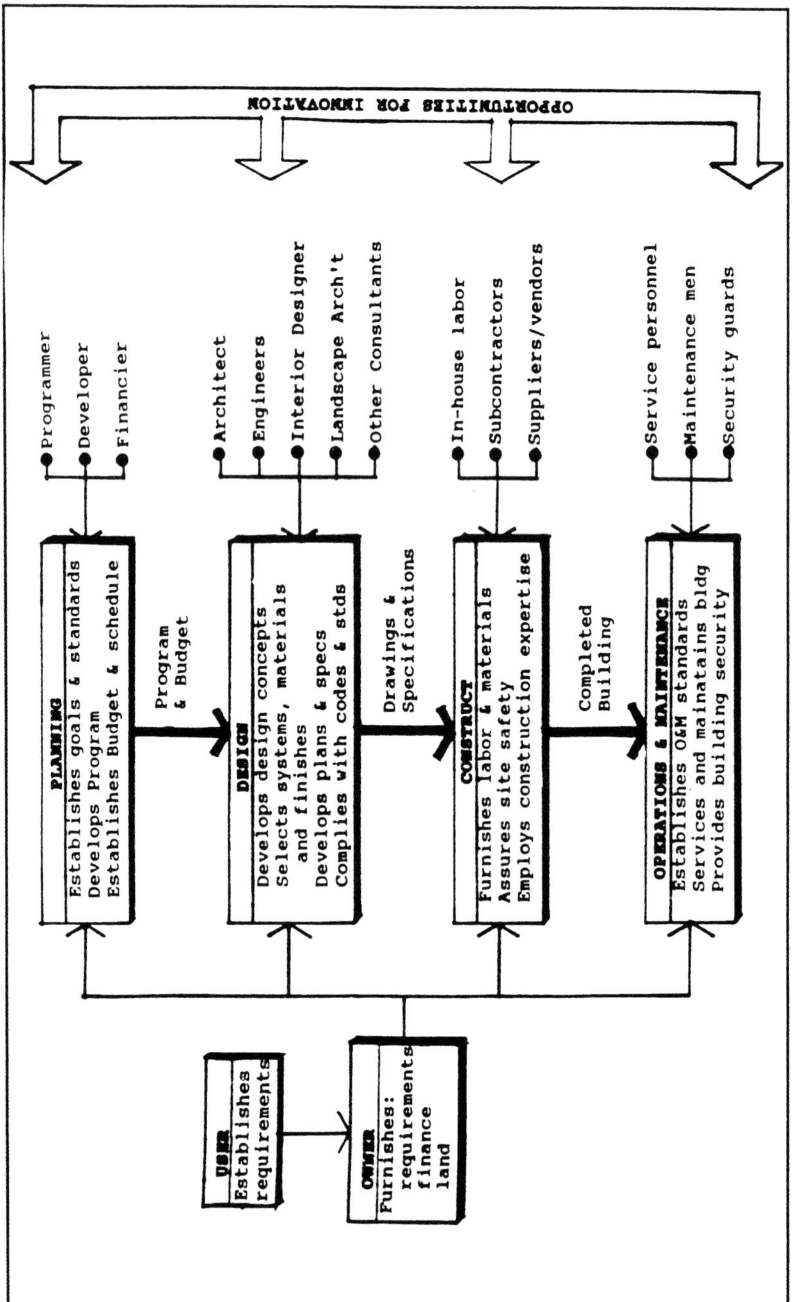

Figure G-1 Overview of design construction process.

The owner's relationship to the facility is, in principle, long term. The "completed project" is not completed in fact until the owner replaces, sells and vacates, or otherwise breaks this relationship. The owner must typically consider both administration (i.e., related to the organization) and operations (i.e., related to the facility and its occupants) in determining what resources are appropriate and how they are employed. Developers of government facilities may contend with budgetary constraints imposed by legislative decisions that have limited relationship to the facility's characteristics.

While operations and maintenance may continue for many years, the greatest management complexity is concentrated in the early stages of the facility's life cycle, during planning, design, and construction. These stages also represent the greatest proportion of potential entry points for new technology. For these early stages, the owner enters into contractual relationships with the designer and the builder (see **Figure G-2**). The contract vehicles contain, in principle, a clear statement of all of the owner's requirements that are to be met during the work leading to the completed project. The contract also states the amount of money the owner agrees to pay for construction. In practice, the requirements may shift during project development, leading to contract changes, negotiations, and sometimes disputes.

The designer is typically a multidisciplinary team of a lead architect or engineering firm and one or more specialist firms that work as subcontractors to the lead firm (see **Figures G-3A** and **G-3B**). The lead design firm generally is completely responsible for selection and direction of the design subcontractors, but many government agencies selecting designers will ask for a list of specialists on the team, prior to making a selection.

The builder (see **Figure G-4**) also is typically a team of a general contractor and one or more specialty subcontractors. The owner, sometimes with advice from the designer, enters into a contract with the general contractor, who is then totally responsible for construction.[100] The owner's requirements, presented in drawings and specifications prepared by the designer, are the basis for this contract.

The construction general contractor will employ specialist subcontractors, and each organization will employ skilled and semiskilled workers. These contractors may have to deal with trade unions that have specific rules regarding utilization of their workers (e.g., trade jurisdiction, work hours, apprenticeship). Each craft's areas of responsibility are often strictly defined, and the contractor must coordinate the work of various specialty workers. The general contractor

---

[100]Some government agencies contract separately with several specialty constructors (e.g., structural, electrical, and mechanical) and must then coordinate these separate contractors. In New York and a few other states, such separate prime contracts are required by law.

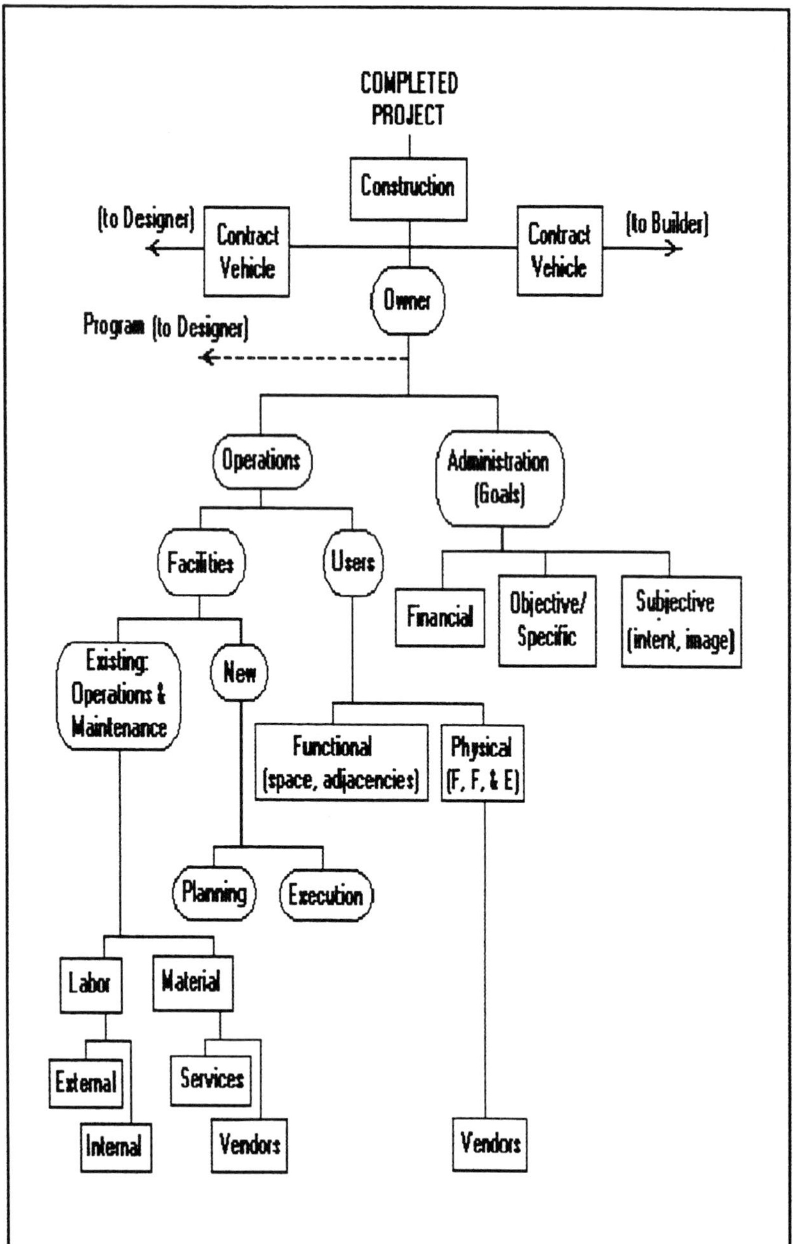

Figure G-2 Detail of design and construction process: owner and contractual relationships.

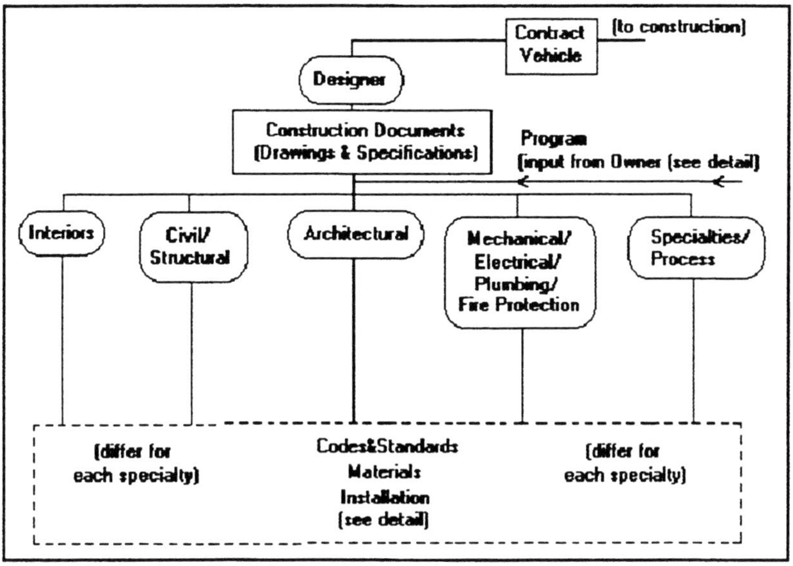

Figure G-3a Detail of design and construction process: designer.

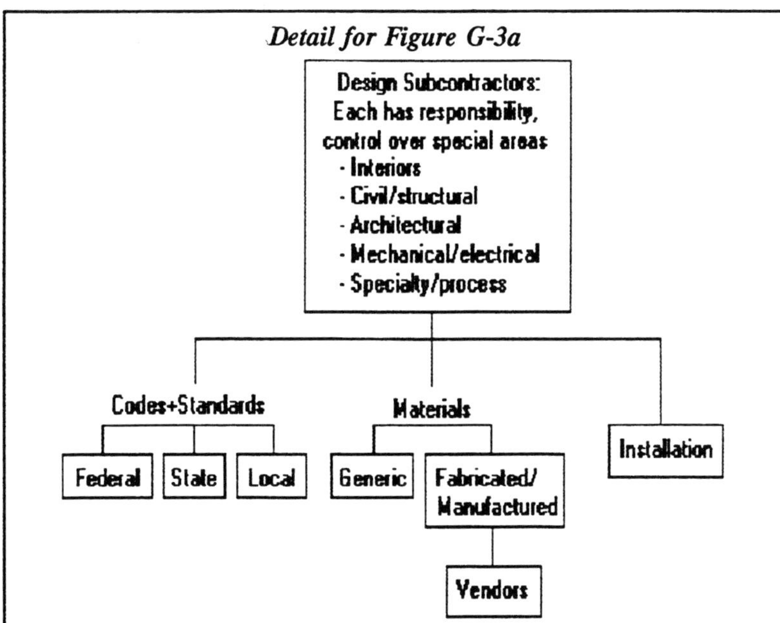

Figure G-3b Primary elements under control of design contractors.

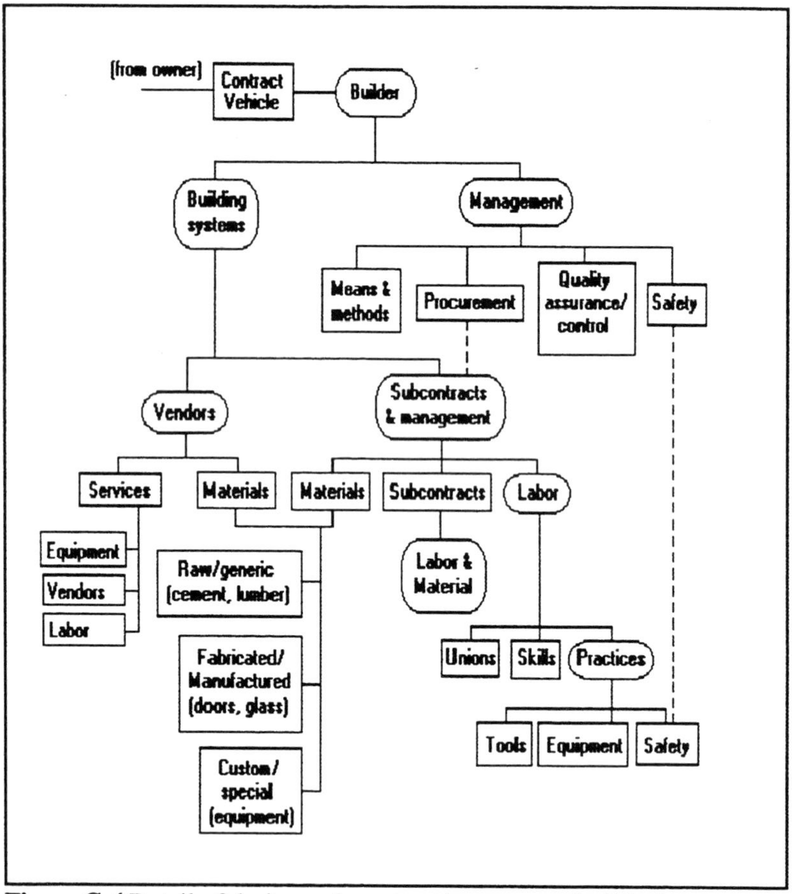

Figure G-4 Detail of design and construction process: builder/constructor.

will also work to ensure that safety is achieved at all levels of the construction work.

New technology may be visualized as entering the process from the lowest levels and moving upward to appear in the final product, the completed project. Innovations can be initiated anywhere in this process, springing from two primary sources: (1) new products, tools, and procedures offered by vendors, and (2) new procedures and relationships initiated by labor, craftsmen, designers, managers, and others working on the project.

Although innovation can start at any point, each of the parties to the process may have particular and differing points of view on the costs and benefits of a proposed new technology (e.g., materials, products, procedures). For example, an improved material may lead to the substitution of that material for another, leading in turn to the loss of sales for some vendors and the loss of work for those who deal with the replaced material. Some participants in the process may oppose new technologies that other participants favor, because the costs and benefits are (or appear to be) distributed disproportionately or simply because the new idea has been proposed by someone else.

New products and processes may then face a tortuous path on the road to becoming innovations. The owner may have great difficulty determining the ultimate value of the potential innovation and may not even have the opportunity to make that judgment. A sophisticated owner may be able to maintain good information on new developments in each of the fields that represent opportunities for innovation, but most owners must generally depend on designers and builders for information and guidance.